THE **BASICS** OF CHEMICAL REACTIONS

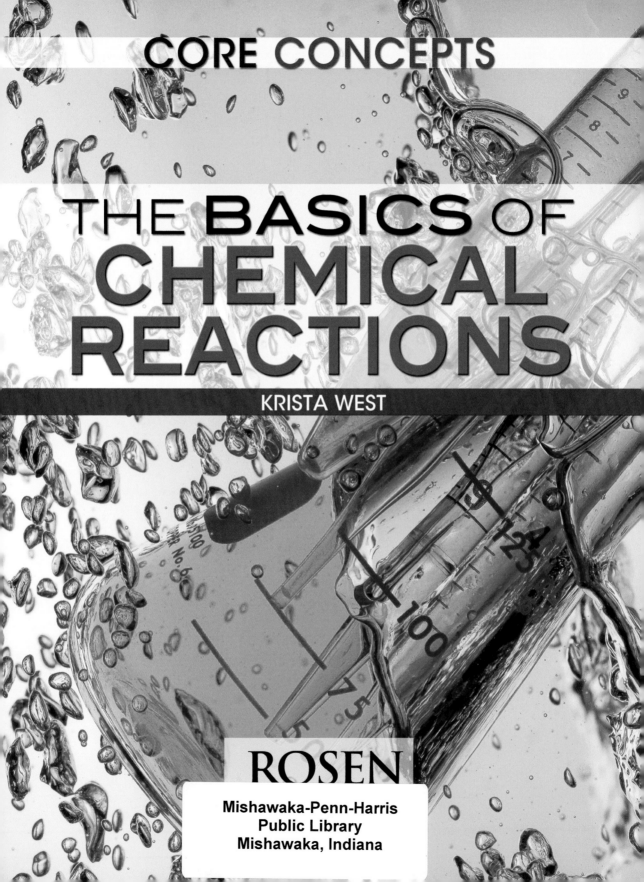

CORE CONCEPTS

THE **BASICS** OF **CHEMICAL REACTIONS**

KRISTA WEST

ROSEN

This edition published in 2014 by:

The Rosen Publishing Group, Inc.
29 East 21st Street
New York, NY 10010

Additional end matter copyright © 2014 by The Rosen Publishing Group, Inc.

Library of Congress Cataloging-in-Publication Data

West, Krista.
The basics of chemical reactions/Krista West.—First edition.
    pages cm.—(Core concepts)
Audience: Grades 7-12.
Includes bibliographical references and index.
ISBN 978-1-4777-2709-6 (library binding)
1. Chemical reactions—Juvenile literature. 2. Chemistry—Juvenile literature. I. Title.
QD501.W6215 2014
541'.39—dc23

2013027821

Manufactured in the United States of America

CPSIA Compliance Information: Batch #WS14YA: For further information, contact Rosen Publishing, New York, New York, at 1-800-237-9932.

© 2007 Brown Bear Books Ltd.

CONTENTS

CHAPTER ONE

CHEMICAL REACTIONS

What turns food into energy, coal into fire, and iron into rust? The answer is chemical reactions. Chemical reactions are taking place all around us and even inside our bodies.

Without chemical reactions, the world would be a very boring place. A chemical reaction is any process that changes one substance into another. Some reactions happen naturally, such as when we digest food or when metal objects become rusty. Other reactions are produced by people to improve their lives. We burn fuel to power an automobile engine, for example.

Reactions involve the interaction between two basic components of the universe—matter and energy. Scientists

All chemical reactions involve change. Burning is one way of converting chemicals such as coal or oil into energy that can be used to power engines or provide heat.

TAKING A CLOSER LOOK

CHEMISTRY AT HOME

Using chemical reactions, chemists have created countless products we use every day. Look around your home and you are bound to see many. Plastics are chains of different types of chemicals strung together. Soaps and toothpastes are made from fatty substances using chemical reactions. And recipes tell us how to use chemical reactions to cook food. Chemical reactions are everywhere.

Dish soap uses a chemical reaction to lift grease molecules from dirty pans and plates.

call any substance that takes up space matter. Rocks, water, and air are all made of matter. Energy is the ability to do work —to move or reshape matter in some way. Heat, light, and electricity are types of energy. During a chemical reaction, energy works to reorganize matter.

ELEMENTS

All matter on Earth is made of elements. An element cannot be reduced to a simpler substance. All elements are made of atoms. An atom is the smallest piece of an element that still has the properties of that element. Atoms do have smaller parts, which have other properties. Chemists represent each element with a symbol of one or two letters.

Atoms are often found in simple combinations called molecules. A pure substance consists of only one type of molecule, which is described by a molecular formula. The formula shows how many atoms of each element are involved. One of the simplest molecules is hydrogen (H_2). This formula shows that the molecule contains two hydrogen (H) atoms. The formula for water is H_2O; two hydrogen atoms are connected to one oxygen (O) atom.

THE INGREDIENTS

The substances you start with in a chemical reaction are called the reactants. The new substances that are created are called the products. Chemists write the reactants and products as chemical equations. All chemical equations follow the same format: Reactants → Products. Numbers are used in the equation to indicate how much of each substance is needed. The arrow indicates that a chemical reaction has taken place and something new has been produced.

A simple chemical reaction occurs when carbon dioxide (CO_2) forms. This molecule contains carbon (C) and oxygen atoms. These two elements combine to produce carbon dioxide. The equation of this reaction looks like this:

$$C + O_2 \rightarrow CO_2$$

A balanced chemical equation shows exactly how much of the reactant and product are involved in the reaction.

KEY DEFINITIONS

• **Atom:** The smallest piece of an element that still retains the properties of that element.

• **Chemical reaction:** A process in which atoms of different elements join together or break apart.

• **Element:** A substance made up of just one type of atom.

• **Matter:** Anything that can be weighed.

Chemists balance equations to determine how many reactants are needed to produce new substances. A balanced equation is one where the number of atoms on one side is the same as the number on the other.

Natural gas is a fuel used in ovens and on cooktops. It releases lots of heat when it burns. The gas is mainly methane, a substance with molecules made of carbon and hydrogen atoms. The burning takes place when methane reacts with oxygen. The products of this reaction are carbon dioxide and water.

The equation for this reaction *(below)* is balanced: the number of each type of atom is the same on both sides.

methane	oxygen	carbon dioxide	water	heat
CH_4	$2O_2$	CO_2	$2H_2O$	

WHAT ARE SUBATOMIC PARTICLES?

The smaller parts of the atom are the pieces involved in a chemical reaction. These parts are called subatomic particles. At the center of the atom is the nucleus. The nucleus is a densely packed ball of positively charged particles called protons. These are mixed with neutral (noncharged) particles called neutrons.

Opposite charges attract, while like charges repel. The positively charged protons in the nucleus attract negatively charged particles called electrons. Electrons are much smaller than protons. They move in clouds around the nucleus. It is the electrons that allow an atom to form bonds with other atoms. How the electrons from two atoms interact determines which type of bond forms. Electrons can be given, taken, or shared to create a bond between two or more atoms. During a chemical reaction, bonds linking some atoms are broken and new bonds are built between others. When atoms of different elements bond they create a substance called a compound. Compounds often look very different than the reactants that produced them. For example, sugar is a compound of carbon, hydrogen, and oxygen. Pure hydrogen and oxygen are both invisible gases, and pure carbon forms diamonds, or graphite, the substance used as pencil lead. Together these elements form many compounds called carbohydrates. These include the sweet-tasting crystals known as sugar.

The products of chemical reactions are often very different from the reactants. Here, a liquid reactant is reacting to produce a solid. Cobalt hydroxide precipitate is formed by adding sodium hydroxide.

THE ENERGY BEHIND REACTIONS

Energy is an essential part of chemical reactions. It is required to break a chemical bond, and energy is released when another bond forms. Heat is one type of energy often involved in chemical reactions. Some reactions will take in heat. Other chemical reactions will give off heat, such as burning fuel.

TOOLS AND TECHNIQUES

A LOOK AT THE PERIODIC TABLE

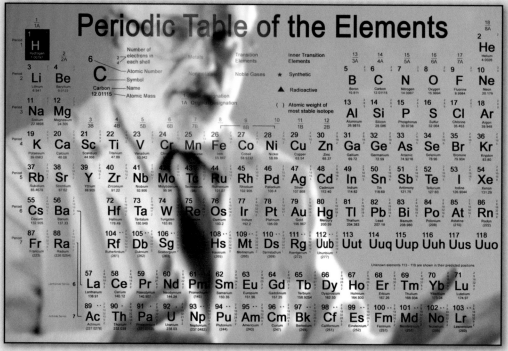

The periodic table of the elements groups together elements with similar properties for easy reference. Chemists and even scholars consult the periodic table from time to time.

Chemists often consult the periodic table, one of the most powerful tools in chemistry. The periodic table is an organized list that provides information about individual elements and groups of elements. The vertical columns in the table are called groups, or families, of elements. Members of each group generally react in the same way.

Each group has a set of known properties. For example, the column on the left of the periodic table is known as the alkali metals. These are very reactive elements, such as sodium and potassium. Instead of memorizing the properties for every element (there are currently 118, with more being discovered all the time), chemists simply consult the periodic table.

TAKING A CLOSER LOOK

THE CHEMISTRY OF LIFE

Without chemical reactions, life could not exist. Like the bodies of all life-forms, the human body is powered by chemical reactions. You inhale oxygen (O_2) when you take a breath of air. When you eat food, your stomach extracts useful chemicals, such as sugar, from it. Oxygen reacts with the sugars in your body to produce carbon dioxide (CO_2) and water (H_2O). Biologists call this chemical reaction respiration. The reaction releases energy from sugar, which keeps the body alive. You exhale the products of respiration with each breath.

Plants complete the same chemical reaction in reverse—a process called photosynthesis. They take in carbon dioxide and water, and use the energy in sunlight to produce oxygen and sugar.

During exercise the body uses the respiration reaction to release energy from food. The energy is used to power other chemical reactions in the muscles that move the body. Plants also use respiration to release energy, but they also do the opposite through photosynthesis. Photosynthesis takes place in the green parts of plants, mostly in the leaves.

REARRANGING THE BONDS

When compounds undergo a chemical reaction, energy works to rearrange the bonds between the atoms. For example, consider the equation: $AB + C \longrightarrow A + BC$.

Elements A and B are bonded to form the AB compound. The AB compound and C are the reactants. During the reaction, the bond between A and B breaks and a bond between B and C is built. A and the BC compound are the products.

In this reaction one bond was broken and a new bond between two different atoms was made. The atoms themselves did not change—A did not change to D, for instance. The reaction changed only how the elements were joined.

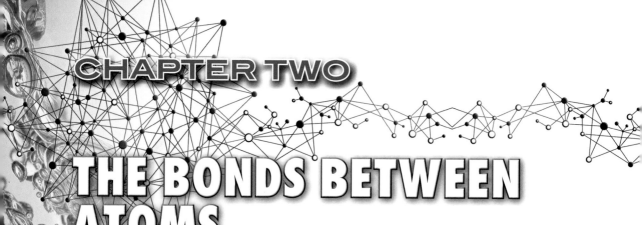

CHAPTER TWO

THE BONDS BETWEEN ATOMS

Chemical bonds allow atoms to stick together in different combinations. How a bond between atoms forms depends on the number and location of the atom's electrons.

Chemical bonds are created when atoms give, take, or share electrons. There are three types of chemical bonds: ionic, covalent, and metallic. The type of bond formed between atoms depends on how many electrons they

An atom's central nucleus contains most of its matter. Electrons move around the nucleus and are the particles involved in chemical reactions.

have in the atom and how they are arranged.

ELECTRONS

The location of electrons in an atom is one factor that determines how that atom will form bonds. Scientists use two models to explain the location of electrons in the atom—the Bohr model and the quantum mechanics model.

The Bohr model describes electrons orbiting (circling) the nucleus of an atom like the planets orbit the Sun. As electrons travel in circles around the nucleus, they are held in place by the pull of the nucleus. The nucleus has a positive charge, which attracts the negative charges of the electrons. This model for the atom works well for very simple atoms, such as hydrogen.

The quantum mechanics model is more modern and mathematical. It describes volumes of space called electron clouds, inside which electrons reside. It is not possible to know exactly where each electron is or how fast it is moving inside a cloud. However, their average positions can be calculated. The quantum mechanics model is a much more complicated and more accurate way of describing how an atom is put together than the Bohr model.

THE IMPORTANCE OF ENERGY LEVELS

In both models, electrons sit at different energy levels. An energy level determines

PROFILE

JOHN DALTON

English chemist John Dalton (1766–1844) is best known for his atomic theory—a basic set of rules that explain how atoms behave, including how they combine with each other. Four of Dalton's rules are still true today. 1) All matter is composed of atoms. 2) All atoms in an element are the same. 3) Atoms combine to form compounds. 4) Atoms are rearranged in a chemical reaction. Only a final rule—that atoms cannot be divided into smaller particles—was wrong. We now know atoms contain even smaller pieces.

how likely an electron is to be involved in a chemical reaction and form a bond. Electrons in energy levels farthest from the nucleus are most likely to become involved in a reaction because they are held only weakly by the nucleus.

Atoms can have several energy levels. The level closest to the nucleus can only hold two electrons. Chemists call this the lowest energy level. The levels farther from the nucleus can hold more than two electrons. Electrons need more energy to sit in the outer energy levels.

The energy levels are sometimes called orbitals—the areas in which electrons orbit around the nucleus. Chemists also describe them as electron shells because they can be thought of as layers, or shells, surrounding the nucleus.

NUMBER OF ELECTIONS

The number of electrons in an atom is the other factor that determines how that atom will react and form bonds. An atom is most stable (unreactive) when its shells full of electrons. An atom with a full shell will not give, take, or share electrons easily. Because of that, the atom does not get involved in chemical reactions and form bonds.

Electrons fill the inner shells first. The lowest-energy shell has space for just two electrons. Helium atoms have two electrons, and these fill the inner shell. This makes the atom stable. It does not give, take, or share electrons because its shell is full.

Larger atoms have two or more electron shells. The extra shells are larger than the first one and need eight electrons to become stable. Becoming stable with eight electrons is called the octet rule. This rule drives all chemical reactions because atoms will react with each other until they become stable.

TAKING A CLOSER LOOK
SOLID, LIQUID, OR GAS

Matter comes in three states—solid, liquid, and gas. Each state depends on the arrangement of the atoms or molecules. Matter changes from one state to another by being heated or cooled. Solids are closely packed groups of atoms or molecules. The atoms and molecules can vibrate back and forth but they cannot move past one another. When a solid is heated it melts into a liquid. A liquid is a collection of atoms or molecules that are less closely packed than in a solid. The molecules have enough spaces between them to flow past each other. Heating a liquid causes it to boil into a gas. A gas is a group of fast-moving atoms or molecules that are completely separate from each other and moving in all directions. A solid has a fixed shape, a liquid takes on the shape of its container, while a gas spreads out and fills all the available space.

The state of matter depends on how hot it is. When the atoms are heated they move more rapidly. This movement begins to break the bonds between them and this results in a change of state.

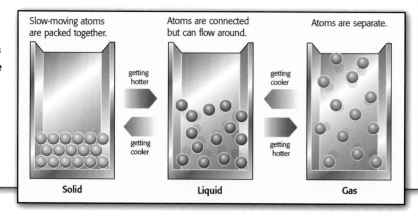

Slow-moving atoms are packed together.

Atoms are connected but can flow around.

Atoms are separate.

getting hotter

getting cooler

getting cooler

getting hotter

Solid   Liquid   Gas

Stable atoms are unreactive. Atoms with incomplete outer shells will give, take, or share electrons to fill their energy levels. These atoms are reactive because they take part in chemical reactions. An atom with one electron in its outer shell will give it away easily. Atoms with six or seven outer electrons readily take electrons to fill that shell.

When atoms give, take, or share electrons they create bonds. During a chemical reaction, bonds are broken and built to create new molecules.

CREATING BONDS

Atoms have a naturally neutral charge. The positive protons in the nucleus balance out the negative electrons. Because of the location and number of electrons, some atoms create bonds more readily

KEY DEFINITIONS

• **Metal:** A hard but flexible element. Metals are good conductors. Their atoms have only a few outer electrons.

• **Metalloid:** An element that has both metallic and nonmetallic properties.

• **Nonmetal:** An element that is not a metal. Nonmetals are poor conductors. Their atoms tend to have several outer electrons.

than others. Based on this ability to form bonds, all elements can be divided into three basic types: metals, nonmetals, and metalloids.

Metal atoms have only a few outer electrons and tend to lose them during chemical reactions. Most elements are metals, and they have certain properties.

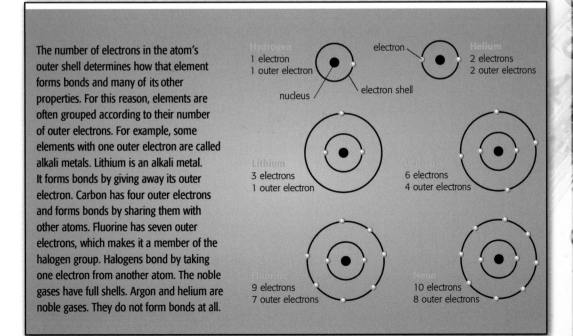

The number of electrons in the atom's outer shell determines how that element forms bonds and many of its other properties. For this reason, elements are often grouped according to their number of outer electrons. For example, some elements with one outer electron are called alkali metals. Lithium is an alkali metal. It forms bonds by giving away its outer electron. Carbon has four outer electrons and forms bonds by sharing them with other atoms. Fluorine has seven outer electrons, which makes it a member of the halogen group. Halogens bond by taking one electron from another atom. The noble gases have full shells. Argon and helium are noble gases. They do not form bonds at all.

Hydrogen
1 electron
1 outer electron

electron

nucleus

electron shell

Helium
2 electrons
2 outer electrons

Lithium
3 electrons
1 outer electron

Carbon
6 electrons
4 outer electrons

Fluorine
9 electrons
7 outer electrons

Neon
10 electrons
8 outer electrons

CHEMISTRY IN ACTION
WHAT IS THE DIFFERENCE?

Iron   Sulphur

When elements combine into a compound they form a completely new substance that has different properties from the original elements. Compounds are not mixtures of those elements. A mixture contains substances that can be separated from each other easily. A compound can only be separated into its ingredients by a chemical reaction that breaks its bonds.

Samples of pure iron and sulfur. When mixed together (*bottom left*) and heated, the iron and sulfur atoms react to form the compound iron sulfide (*bottom right*).

Metals are solid, shiny, and can conduct electricity. A solid piece of metal contains many free-floating electrons shared among the atoms. These electrons act like stepping stones for an electric charge within the metal, allowing the charge to move within the solid. This property makes metals good materials to use in wires and electric cables.

Nonmetals are the opposite of metals. Nonmetals tend to gain electrons in chemical reactions. They come in all forms and may be liquid, gas, or solid in normal conditions. They are not good at conducting electricity. There are no free-floating electrons to help move an electric charge.

Instead, nonmetals are good insulators. Metalloids are semiconductors, which change from being insulators to conductors depending on the conditions.

How strongly an atom holds onto its electrons and pulls electrons away from another atom is called its electronegativity. Nonmetals are more electronegative than metals. Metal atoms have only a few outer electrons, which the atoms give away easily.

WHAT IS AN IONIC BOND?

During a chemical reaction, three types of bonds can form: ionic, covalent, or

Salt is found in desert salt flats and dissolved in seawater

metallic. An ionic bond occurs when a metal atom gives an electron to a nonmetal atom. The atom that gives away an electron loses a negative charge and becomes positively charged itself.

Chemists call atoms that have become charged in this way ions. A positively charged ion is a cation. The atom that takes an electron receives an extra negative charge and becomes a negatively charged ion, or anion. Opposite charges attract each other, so the cation and anion join to form an ionic bond. Bonded ions are called ionic compounds.

A common ionic compound is table salt. Table salt is formed when sodium (Na) bonds with chlorine (Cl). Sodium is a typical metal. It is silvery and conductive, with one electron to give. Chlorine is a nonmetal gas, in need of one electron to complete its octet and become stable.

KEY DEFINITIONS

• **Conductor:** A substance that carries electricity and heat well.

• **Electricity:** A stream of electrons or other charged particles moving through a substance.

• **Insulator:** A substance that does not transfer an electric current or heat.

Put sodium and chlorine in a container together, and sodium will lose its electron and become a cation (Na+). Chlorine takes the same electron and becomes an anion (Cl-). The Na+ bonds with the Cl- to form NaCl—sodium chloride, or the table salt we use in food.

All ionic molecules are formed by cations and anions bonding. This gives the molecules a positively charged end, or pole, and a negatively charged pole. Each pole is attracted to another with the opposite charge on a different molecule. As a result, ionic molecules tend to join up in regular patterns, called crystals. Because each molecule is held firmly in place by all the other molecules around it, ionic crystals tend to be hard solids that do not bend or break easily.

Because of the strength of the attraction between ions, it takes a lot of energy to pull them apart. Heating the solid provides enough energy to pull some of the molecules apart and they melt into a liquid. The temperature at which a substance melts is called its melting point. Further heating separates the molecules more until the liquid boils to produce a gas. The temperature when this happens is called the boiling point. Ionic compounds tend to have high melting and boiling points.

When an ionic compound is dissolved in water, the ions separate and float freely in the water. These floating ions can carry electricity through the water. The ability to conduct electricity when dissolved or melted is another common property of ionic compounds.

TOOLS AND TECHNIQUES

CHARTING ELECTRONS

Chemists show how atoms form bonds by drawing atoms in a simple way. The nucleus is a central circle and is surrounded by layers of electrons. During chemical reactions electrons move between atoms or are shared in the outer shells of two atoms. The diagram below shows how an electron moves from a sodium atom to a chlorine atom to make sodium and chloride ions. The ions bond together to make sodium chloride, or common salt.

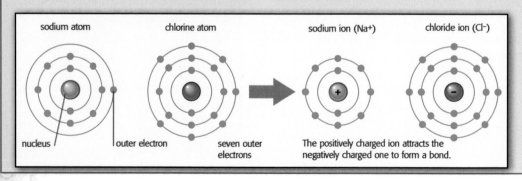

sodium atom   chlorine atom   sodium ion (Na+)   chloride ion (Cl-)

nucleus   outer electron   seven outer electrons   The positively charged ion attracts the negatively charged one to form a bond.

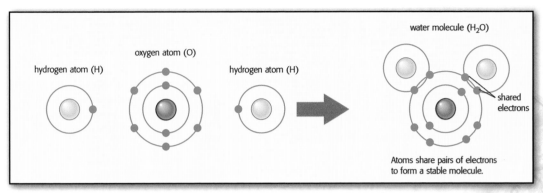

water molecule (H_2O)

hydrogen atom (H)

oxygen atom (O)

hydrogen atom (H)

shared electrons

Atoms share pairs of electrons to form a stable molecule.

WHAT IS A COVALENT BOND?

A covalent bond is formed when two nonmetal atoms share their electrons to become stable. Instead of one electron moving into the outer shell of another atom, the shells overlap to share the electron. Each electron becomes bonded to both the nuclei.

A group of atoms held together with bonds like this is called a covalent molecule. Hydrogen atoms form the simplest covalent molecules. Hydrogen has only a

Water is a covalent compound. Two hydrogen atoms share electrons with one oxygen atom.

single electron and only needs two electrons to become stable (instead of eight). A hydrogen atom shares its electron with another hydrogen atom, forming a H_2 molecule. Hydrogen is found in nature as H_2. Six other elements form molecules in a similar way: oxygen (O_2), nitrogen (N_2), fluorine (F_2), chlorine (Cl_2), bromine (Br_2), and iodine (I_2).

Sand (silicon dioxide) and water are two of the most common compounds on the surface of Earth. Both are covalent compounds. However, silicon dioxide forms hard crystals, while water is a liquid.

TRY THIS

CREATING RUST

Place an iron nail in a jar. Cover the nail completely with water and add two tablespoons of salt. Put the lid on the jar. Come back and check the nail in about an hour. What do you see? You are watching a chemical reaction in action:

iron + oxygen → iron oxide

A new ionic bond between iron and oxygen forms to create iron oxide, or rust. Salt and water help the reaction go faster. The rust should look like dark, reddish spots on the nail.

When it gets wet, iron reacts with oxyen to form rust, a red, flaky iron oxide.

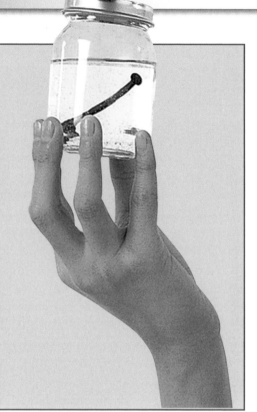

Because all covalent bonds involve the sharing of electrons, covalent compounds tend to have similar properties. The crystals of covalent compounds fall into two types. The first type is similar to an ionic crystal because all the atoms are connected to each other by a strong bond. This is the case with diamond, which is an extremely hard form of pure carbon. Diamond is the hardest substance known. Silicon dioxide, which is the scientific name for sand and quartz, also forms hard crystals in this way. Both diamond and silicon dioxide have high melting and boiling points because of the strong network of bonds inside their crystals.

The other type of covalent compounds do not form crystals in the same way. They have no strong forces to bond molecules to each other. Instead only weak forces, called van der Waals forces, pull molecules together. Because these forces are so weak, solids are only formed at very cold temperatures. In normal conditions, the covalent compounds are gases or liquids. For example, carbon dioxide is normally a gas and will only form a crystal at very low temperatures.

Because electrons are shared in covalent compounds, there are no free-floating charged particles to conduct electricity. As a result, covalent compounds tend to be good insulators.

Copper (*left*) and zinc (*center*) are metal elements. They can be mixed together to form an alloy called brass (*right*). Inside the alloy, the copper and zinc atoms are joined by metallic bonds.

Covalent compounds that include carbon (C) and hydrogen (H) are called organic compounds. Many burn easily when exposed to oxygen and are used as fuels. Gasoline, for example, is a mixture of several organic compounds. The term *organic* is used for these compounds because many were originally made by living organisms.

WHAT ARE METALLIC BONDS?

Metals are generally hard solids. They tend to be flexible, too. The atoms are held together by metallic bonds, and a metal's properties are a result of these bonds. Metallic bonds occur when metal atoms share a pool of electrons. Unlike a covalent bond, where electrons

Metallic bonds make it possible for metals, such as copper, to be pulled into long, thin wires. This is called being ductile. Metals are also malleable—they can be flattened into thin, flexible sheets that bend but do not break easily.

are shared but still bound to a nucleus, electrons in a metallic bond are free to move around.

A piece of solid silver consists of atoms floating in a pool of free electrons. All the metal atoms give away their single outer electron, which make up a negatively charged pool of electrons. This negative charge is attracted

KEY DEFINITIONS

• **Covalent bond:** A bond in which two or more atoms share electrons.

• **Ionic bond:** A bond produced when oppositely charged ions are attracted to each other.

• **Metallic bond:** A bond between a group of metal atoms that are sharing a pool of electrons.

• **van de Waals bond:** A very weak bond that attracts molecules to each other.

TRY THIS

PENNY EXPERIMENT

Sand the edge of a penny so the outer layer of copper is rubbed away exposing zinc underneath. Place the penny in a pint of vinegar for one hour. Now remove the penny and add 1.5 ounces (50 g) of Epsom salts and 2 ounces (60 g) of sugar to the vinegar. Use alligator clips to attach electrical wires to the sanded penny and a new, clean penny. Lower the pennies into the vinegar, making sure they do not touch. Connect the clean penny to the negative terminal of a large battery. Connect the sanded penny to the positive terminal. After 10 minutes, the clean penny should have a dark, silvery coating of zinc on it.

You used electricity to break the metallic bonds between copper and zinc on the sanded penny. The vinegar, salts, and sugar helped move the zinc from the positively charged penny to the negatively charged penny. Scientists call this electroplating.

A new penny is bright and shiny. But after a few years of regular use, it will be dark like the older pennies shown here.

An electric current flows between the two pennies through the mixture of vinegar, salts, and sugar. The zinc is carried by this current from the sanded penny to the other coin.

to the positive charge of the atoms' nuclei. This attraction bonds the metal atoms together.

Although metallic bonds make most metals hard solids, they also allow the atoms inside to move past each other. This property is what makes metals ductile and malleable. Ductile solids can be pulled into thin wires, and malleable ones are easily flattened into sheets. The pool of electrons holds the metal atoms together as the solid is reshaped and prevents the solid from breaking.

WHAT ARE ISOMERS?

Isomers are molecules that contain the same atoms but are arranged in different ways. They have the same chemical

TAKING A CLOSER LOOK
ISOMER STRUCTURES

The compound C_3H_8O has three forms, or isomers. The isomers have different chemical properties. Two of them are types of alcohols while the other is an ether.

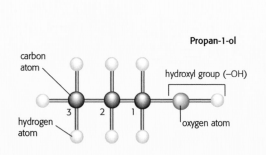

▲ This alcohol molecule is called propan-1-ol. The hydroxyl group, which all alcohols have, is attached to the first carbon (1) atom.

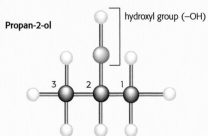

▲ This alcohol molecule is called propan-2-ol because the hydroxyl is bonded to the second carbon (2) atom.

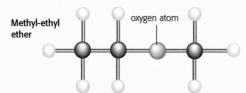

▲ The third isomer is not an alcohol. Instead the oxygen atom is bonded to two carbon atoms. Molecules like this are called ethers.

formula but have a different shape. Understanding isomers is important in chemical reactions because isomers have different atoms available for bonding.

Good examples of isomers are the organic compounds with the formula C_3H_8O. These molecules take three forms. Two are a type of alcohol and are known as propanols. Alcohols are a group of organic compounds. Alcoholic drinks contain an alcohol called ethanol. However, propanol and all other alcohols are extremely poisonous. In propanol molecules, the oxygen atom is bonded to any of the three carbon atoms and to a hydrogen atom. Together the hydrogen and oxygen form a hydroxyl group (–OH). All alcohols have a hydroxyl group, which plays a role in chemical reactions. The two C_3H_8O alcohols are called propan-1-ol and propan-2-ol.

In the third C_3H_8O molecule the oxygen bonds between two carbons. This compound is not an alcohol. Instead it is an ether called methyl-ethyl ether. Ethers react in a different way to alcohols.

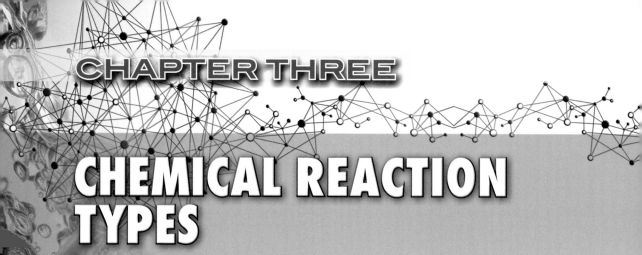

CHAPTER THREE

CHEMICAL REACTION TYPES

Chemists classify chemical reactions according to how chemical bonds are broken or built. Then the scientists write those reactions as chemical equations.

Chemical reactions occur when new bonds are formed between atoms to create new compounds. Chemists have names for many different types of chemical reactions. The type of chemical reaction depends on how the reactants change to make the products. The five main types of chemical reactions are combination, decomposition, displacement, redox, and combustion reactions. Remember these are general groups, and it is

Chemical reactions can involve reactants and products in all states. For example, reacting liquids can produce gases.

possible for a chemical reaction to be a member of more than one group at the same time.

Some elements are involved in nuclear reactions. These are very different from chemical reactions. Instead of rearranging the bonds to form new compounds, a nuclear reaction actually changes the atom, making one element into another.

WHAT ARE COMBINATION REACTIONS?

A combination reaction occurs when two or more reactants combine to form one product: $A + B \longrightarrow AB$. In more complex cases where there are several reactants, more than one product can be formed by this type of reaction. Common compounds, such as water,

PROFILE

ROBERT BOYLE

Robert Boyle (1627–1691) was one of the first chemists. His work helped future scientists figure out what was happening during chemical reactions. Boyle was born in Ireland but lived in England. He became interested in chemicals as he looked for a way of turning common metals into gold. He failed but learned several things in the process. In 1661 Boyle published *The Skeptical Chemist*, in which he suggested that matter was made of many elements. Before this people thought the only elements were earth, wind, water, and fire. Boyle also showed how gases changed as they were heated and squeezed. A hot gas takes up more room than a cold gas. In addition, when a gas is squeezed it gets hotter. This relationship is called Boyle's law. The law helped later chemists understand what gases are made of.

Robert Boyle and Denis Papin in Boyle's laboratory in London, England. The laboratory contains furnaces and flasks used to heat reactants and collect the products.

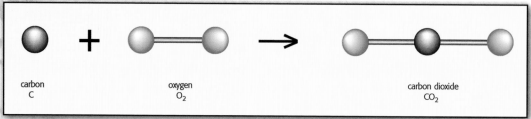

carbon
C

oxygen
O_2

carbon dioxide
CO_2

The diagram above shows a combination reaction between carbon (C) and oxygen (O_2). These reactants combine to form carbon dioxide gas (CO_2).

This is the reaction that takes place when charcoal burns. Charcoal is mainly carbon.

carbon dioxide, and common salt are all the products of combination reactions.

When carbon burns it is taking part in a simple combination reaction. Oxygen combines with the carbon, producing carbon dioxide gas and some heat. The chemical equation representing this is:

$$C + O_2 \rightarrow CO_2$$

This is also an example of a combustion reaction and a redox reaction.

WHAT ARE DECOMPOSITION REACTIONS?

A decomposition reaction is the opposite of a combination reaction. It occurs when a single compound breaks down into two or more simpler substances: $A + B \rightarrow AB$

When you open a can of soda, carbon dioxide bubbles are formed by a decomposition reaction. The soda contains carbonic acid (H_2CO_3) dissolved in water. This mixture is squeezed inside a soda can under high pressure. When the can is opened, the pressure inside drops. This causes the acid to decompose and form water and bubbles of carbon dioxide gas that give the soda its refreshing taste. The decomposition reaction looks like this:

$$H_2CO_3 \rightarrow H_2O + CO_2$$

WHAT ARE DISPLACEMENT REACTIONS?

A displacement reaction occurs when a more reactive atom replaces a less active atom in a compound: A + BC → AB + C. A reactive atom is one that forms bonds easily. Chemists divide displacement reactions into two types: single and double displacements.

A single displacement reaction occurs when the reactive atom doing the replacing is an element. A double displacement occurs when the atom doing the replacing is already combined in another compound.

Double displacement reactions often happen in solutions. A solution is a mixture in which a solid is spread out evenly in a liquid until it has disappeared. This process is known as dissolving. Seawater is a solution of salt.

Double displacements are either precipitation reactions or neutralization reactions. In a precipitation reaction, one product is a compound that cannot dissolve. Instead it forms a precipitate, a solid that is separate from the solution. Precipitates eventually sink to the bottom of the solution.

In a neutralization one of the products is always water. The compounds that undergo neutralization reactions are called acids and bases. Acids are compounds that contain hydrogen ions (H+). Bases contain negative hydroxide ions (OH-). When acids and bases react they form water and another compound. This second product is neither an acid nor base and is described as neutral.

A neutralization occurs when you add household ammonia (a base; NH_4OH) to vinegar (an acid; CH_3O_2H). The

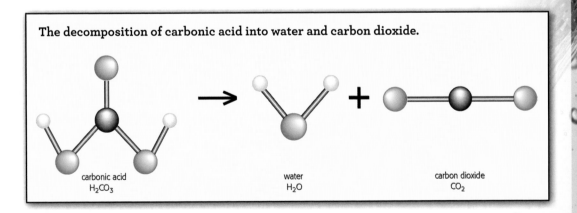

The decomposition of carbonic acid into water and carbon dioxide.

carbonic acid
H_2CO_3

water
H_2O

carbon dioxide
CO_2

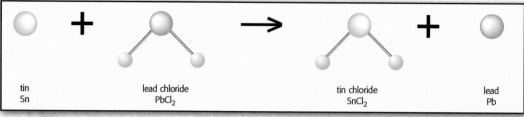

tin
Sn

lead chloride
$PbCl_2$

tin chloride
$SnCl_2$

lead
Pb

The diagram above shows how tin displaces lead when added to a solution of lead chloride.

vinegar's H+ ions and the ammonia's OH- ions combine to form water (H_2O). The NH_4^+ ion bonds with the vinegar ion ($CH_3CO_2^-$). This forms ammonium ethanoate ($NH_4CH_3CO_2$), a neutral compound.

WHAT ARE REDOX REACTIONS?

A redox reaction occurs when electrons move from one of the reactants to the other. Combination, combustion, and single displacement reactions are also

KEY DEFINITIONS

• **Dissolve:** To form a solution.

• **Insoluble:** A substance that cannot dissolve.

• **Precipitate:** An insoluble solid formed by a double displacement reaction between two dissolved compounds.

• **Solute:** The substance that dissolves in a solvent.

• **Solution:** A mixture that contains a dissolved substance. Solids usually dissolve in liquids.

• **Solvent:** The liquid that solutes dissolve in.

This picture shows a displacement reaction in progress. The silver-colored strip of zinc dissolves in the blue copper sulfate solution. Red-brown copper metal forms at the bottom of the test tube.

considered redox reactions. The word *redox* is short for "reduction-oxidation." Each redox reaction involves two separate reactions that occur at the same time. One half of the reaction occurs when one compound gains electrons. Chemists say an atom that has gained electrons has been reduced. The other half of this

TRY THIS
ACID OR BASE?

Chemists use a substance called an indicator to test if something is an acid or a base. The indicator changes color when acid or bases are added to it. You can make an indicator at home from red cabbage.

Chop up a whole red cabbage into small pieces. Boil the pieces for 30 minutes. (Ask an adult to help you and be careful with the hot water.) The boiling cabbage will make the water turn red. Let the water cool and then use a sieve (strainer) to separate the cabbage from the water.

Put the red water in two cups. Add a teaspoon of baking soda to one of the cups. Baking soda is a base. Add a teaspoon of vinegar to the other cup. Vinegar is an acid. What colors do you see? Try testing other substances to see if they are acids or bases.

Boiling red cabbage makes a substance called anthocyanin mix into the water. The color of the anthocyanin depends on how many hydrogen ions are also in the mixture. Acids have many hydrogen ions, while bases do not have any.

acid   base

Red-cabbage indicator turns red when an acid is added and purple with a base.

The color of an indicator depends on how many hydrogen ions (H^+) are present. Chemists measure the amount of H^+ in a solution as its pH. This stands for "potential hydrogen." A solution with a low pH has a lot of hydrogen ions in it. Acids have a low pH. Bases have a high pH. Instead of H^+ ions, they have a large number of hydroxide ions (OH^-). When OH^- and H^+ ions meet they combine into water (H_2O). Water is neutral—it is not an acid or a base. It has a pH of 7. Anything with a pH below 7 is an acid. Bases have a pH of more than 7.

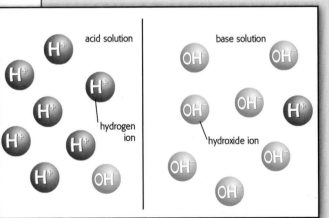

acid solution — hydrogen ion (H^+)

base solution — hydroxide ion (OH^-)

Zinc reacts with an acid. The products of this reaction are a zinc compound and pure hydrogen gas. The gas bubbles out of the liquid. This is both a displacement and a redox reaction. The zinc has displaced the hydrogen from the acid. It has also lost electrons and been oxidized. The acid has gained electrons and been reduced.

reaction, which happens at the same time, occurs when another compound loses electrons. Chemists say an atom that has lost electrons has been oxidized.

The burning of hydrogen is a redox reaction. The two halves of the reactions are written with the following equations:

$$2H_2 \rightarrow 4H^+ + 4 \text{ electrons}$$
$$O_2 + 4 \text{ electrons} \rightarrow 2O^{2-}$$

The two halves are combined into:

$$2H_2 + O_2 \rightarrow 2H_2O$$

The hydrogens have given their electrons to the oxygens and have been oxidized. The oxygens have gained electrons from the hydrogens and have been reduced.

WHAT ARE COMBUSTION REACTIONS?

A combustion reaction occurs when a compound reacts with the oxygen in the air and burns, producing flames and heat. Combustion reactions are often used by people to release heat.

A diagram showing a very simple redox reaction: hydrogen burning with oxygen to form water. The hydrogen is oxidized while the oxygen is reduced.

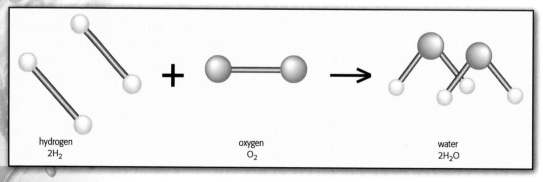

hydrogen
$2H_2$

oxygen
O_2

water
$2H_2O$

CHEMISTRY IN ACTION
REDOX REACTIONS

Redox reactions are everywhere in nature, and they are used a lot by people, too, especially to purify elements.

Photosynthesis and respiration are both redox reactions. Another takes place when when you slice open an apple. Once exposed to the air, the inside of the fruit turns brown in a short time. This happens because substances in the fruit are oxidized by oxygen in the air. They then form a brown compound called melanin.

Redox reactions are also used to purify some metals. Metals are found in nature as part of compounds called ores. A redox reaction is used to remove the metal atoms from the ore. The most common reaction of this type is smelting. Smelting is used mainly to purify iron. During smelting iron ore is heated with carbon. The carbon reacts with the ore and forms a compound with the atoms that were previously bonded to the iron. This leaves pure iron behind. The iron ore has been reduced while the carbon has been oxidized.

Apples turn brown due to redox reactions.

Molten (liquid) iron is poured into a cast after being purified in a smelter. The smelter uses a redox reaction between carbon and iron ore.

TRY THIS

CREATING PHOTOSYNTHESIS

Photosynthesis is a redox reaction that occurs inside plants. You can watch it happening with a simple activity. Place a small piece of pondweed in a glass jar full of water. Cover the jar with a saucer and, holding them together, turn them upside down. Quickly fill the saucer with water to stop water leaking from the jar. A small amount of air should be trapped in the jar. Mark the level of the water with a pen. Now put the jar in a sunny place.

Soon there will be bubbles on the plant and the water level may have dropped slightly. The plant is using sunlight to react water with carbon dioxide to make oxygen and sugar. The sugar is the plant's food.

Photosynthesis produces bubbles of oxygen, which increase the amount of gas in the jar.

Compounds known as hydrocarbons are often burned in combustion reactions. The name hydrocarbon is used because the compounds are made from carbon (C) and hydrogen (H).

For example, propane gas (C_3H_8) is burned in cooking stoves to produce heat for cooking food. The reaction looks like this:

$$C_3H_8 + 5O_2 \rightarrow 3CO_2 + 4H_2O$$

Hydrocarbon fuels are extracted from petroleum oil and natural gas. They are

A diagram showing the combustion (burning) of propane.

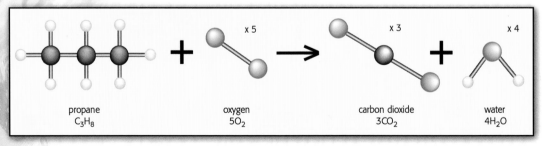

| propane C_3H_8 | oxygen $5O_2$ | carbon dioxide $3CO_2$ | water $4H_2O$ |

often called fossil fuels because they are the remains of plants and other living things buried millions of years ago.

WRITING OUT REACTIONS

You already know that chemists write chemical equations to describe what happens during a chemical reaction. Chemical equations can be written simply, with only the most necessary information included. They can also be written in detail, giving much more information about what happens and what is needed for a reaction to happen.

A simple example of a chemical equation is when two hydrogen atoms combine: $H + H \longrightarrow H_2$. This equation shows that two hydrogen atoms (H) combine into a hydrogen molecule (H_2).

A more detailed chemical equation might include other symbols, letters, and numbers. Much of this extra information added by these symbols is obvious to chemists or is not vital, so such symbols are not always included in equations. However the symbols can add useful information about how to perform the reaction successfully. A vertical arrow pointing up indicates that a product will form a gas that will bubble out of solution. Carbon dioxide gas is a product of several decomposition reactions and is indicated like this: $CO_2\uparrow$. An arrow pointing down

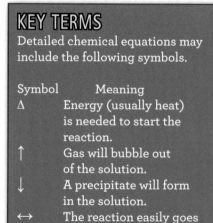

KEY TERMS
Detailed chemical equations may include the following symbols.

Symbol	Meaning
Δ	Energy (usually heat) is needed to start the reaction.
$\uparrow$	Gas will bubble out of the solution.
$\downarrow$	A precipitate will form in the solution.
$\leftrightarrow$	The reaction easily goes both directions.
(s)	The compound is solid.
(l)	The compound is liquid.
(g)	The compound is a gas.
(aq)	The compound is aqueous, or dissolved in water.

A mixture of propane and butane gas explodes in a ball of fire. Explosions are combustion reactions that occur very quickly and release large amounts of heat and gas all at once.

TAKING A CLOSER LOOK

A BALANCE

When a reaction occurs in both directions at the same speed, a dynamic equilibrium is produced. In this system, reactants are combining to form a product, but at the same time, the product is splitting up to form the reactants again. These two processes happen at the same speed, so although reactions are always taking place, the amounts of reactants and products stays the same. Hydrogen (H) and iodine (I) react to form hydrogen iodide (HI) in a dynamic equilibrium. The reaction is represented with the equation:

$$H_2 + I_2 \leftrightarrow 2HI$$

In a detailed chemical equation letters in parentheses are used to indicate the state of matter of each compound. A solid is indicated with (s), a liquid is indicated with (l), and a gas is indicated with (g). When a chemical is dissolved in water, it is described as aqueous (from *aqua* the Latin for "water") and is indicated with (aq).

Sometimes it is easy to predict the states of matter involved in an equation, but not always. Water is usually in liquid form: $H_2O(l)$. However, water involved in reactions with a lot of heat often turns into a gas: $H_2O(g)$. While symbols

indicates that a product will form as a precipitate and sink to the bottom of a liquid. Metals, including silver (Ag), often precipitate in displacement reactions and are indicated like this: Ag↓.

A two-headed arrow between the reactants and the products indicates that the reaction can go in both directions easily. This means that the reactants form the products and the products can also form the reactants. Instead of writing two equations:
A + B ⟶ AB and AB ⟶ A + B, chemists combine the two: A + B ↔ AB.

A student uses a method called titration to measure how much of a reactant is present in a mixture. A second reactant is poured into the other using a measuring tube called a burette.

and states of matter are often helpful pieces of a chemical equation, the numbers are the most important to chemists.

THE NUMBERS

A balanced equation tells you how much of the reactants are needed to produce a certain amount of the products. Chemists alter the amount of a reactant added to a reaction to predict how much of a product will be created. When you read a chemical equation, you see two types of numbers, but only one of these can be changed by the chemist.

The small number to the lower right of an element indicates how many atoms are needed. For example, the "2" in H_2O shows that there are two hydrogen atoms in a molecule of water. There is also one oxygen atom but the "1" is never written. If you do not see a number, you can assume there is only one atom. Chemists can never change the small numbers in a compound in order to balance an equation, because that would imply different bonds between the atoms.

The larger number to the left of a compound is called a coefficient. The coefficient tells you how many of those molecules are needed. For example, $3H_2O$ means there are three water molecules taking part in the reaction. You can multiply the coefficient by the number of atoms in each molecule to count the total number of atoms. In this case there are six hydrogen atoms (3 × 2) and three oxygen atoms (3 × 1).

TOOL AND TECHNIQUES
BALANCING TIPS

If you get stuck balancing a complicated chemical equation, try putting a coefficient of 2 in front of the most complicated compound in the reaction. Then work to make the other atoms match. If this does not work, try again with a coefficient of 3. Keep increasing this coefficient until you find the balance.

BALANCING THE EQUATION

Chemists balance equations by changing the coefficients so that the number of atoms in the reactants equals the number in the products. For example, the equation for making hydrogen molecules, $H + H \rightarrow H_2$, is balanced because the number of atoms is the same on both

KEY DEFINITIONS

• **Chemical equation:** Symbols and numbers that show how reactants change into products during a reaction.

• **Chemical formula:** A combination of chemical symbols that shows the type and number of elements in a molecule. H_2O is the formula for water, which contains two hydrogen (H) atoms and one oxygen (O).

• **Chemical symbol:** Letters used to represent a certain element, such as O for oxygen or Na for sodium.

• **Coefficient:** A number placed in front of a chemical formula to show how many molecules are used or produced by a reaction. $3H_2O$ stands for three water molecules.

sides. However the equation for reacting hydrogen with oxygen to make water is more complex. Both hydrogen and oxygen exist as molecules of two atoms: H2 and O_2. Water contains atoms of both elements but the simple equation $H_2 + O_2 \rightarrow H_2O$ does not balance. There is one oxygen atom in the product, but two oxygen atoms used as the reactants.

To balance this equation, you would add a coefficient to the reactants to match the number of atoms in the products: $2H_2 + O_2 \rightarrow 2H_2O$. Now the number of atoms is the same in the reactants and the products, so the equation is balanced.

WHAT IS A BALANCING TABLE?

The chemical reaction to make water is fairly simple, so balancing the equation is easy. When equations get more complicated, it can be helpful to draw a table to help you balance numbers in the reactants and products. The table lists the number of atoms of each element. Here is a table for $2H_2 + O_2 \rightarrow 2H_2O$:

	Reactants	Products
H	2 × 2 = 4	2 × 2 = 4
O	1 × 2 = 2	2 × 1 = 2

It is clear that you have a balanced equation because there are four atoms of hydrogen and two atoms of oxygen in both the reactants and the products.

When you balance an equation, you add or change coefficients until the numbers of atoms on both sides of the table match.

Now try to balance this equation of the reaction between phosphorus (P_4) and oxygen (O_2):

$$P_4 + O_2 \rightarrow P_2O_5$$

This equation has an unbalanced number of atoms.

	Reactants	Products
P	4	2
O	2	5

To begin balancing the equation, look at the most complicated compound (P_2O_5) and add a coefficient of 2. This will give you four phosphorus atoms and ten oxygen atoms in the products.

Now balance the reactants. Oxygen molecules (O_2) each have two atoms in them, so five molecules would be needed to have ten atoms. Add a coefficient of 5 to the oxygen reactants. There are already four phosphorus atoms in the reactants, so you have finished. Using these numbers, the balanced equation would be:

$$P_4 + 5O_2 \rightarrow 2P_2O_5$$

The number of atoms is the same in the reactants and products, so the equation is balanced.

THE REALITY

With a balanced equation, a chemist knows how many reactants are needed to create the products. However, you cannot count the number of atoms easily in a laboratory. Instead, chemists calculate the number of atoms in a substance by weighing it.

Atoms are counted in moles. The word *mole* represents a number, just as a *pair* equals two, a *dozen* is equal to 12, and a *gross* is 144. A mole is equal to 602,213,670,000,000,000,000,000 (6.022×10^{23}). This number represents the number of atoms or molecules in a set weight of a compound. The number is called Avogadro's number. It is named for Italian Amedeo Avogadro (1776–1856), who discovered that a set amount of any gas always contains the same number of atoms or molecules. On the periodic table, an element's atomic mass number is equal to the number of grams a mole of that element weighs. For example, helium has an atomic mass of 4. This means that one mole of helium weighs 4 grams (0.14 ounces).

Compounds have molecular masses. That is the sum of all the atomic masses of the atoms inside a molecule. For example, the molecular mass of sodium chloride is roughly 58.5—23 for sodium and 35.5 for chlorine. Therefore a mole of sodium chloride weighs 58.5 grams (2 ounces).

Chemists use these weights to measure exactly how much of an element or compound has been used up or produced in a reaction. That is the best way to figure out how the atoms have recombined to form compounds.

Giant balloons float down Broadway in New York City as part of a Thanksgiving parade. The balloons float because they contain helium gas. Helium atoms are smaller and lighter than the atoms in the air. A mole of helium contains the same number of atoms as a mole of another element but weighs much less.

CHAPTER FOUR

USING AND CREATING ENERGY

Chemical reactions can release or absorb energy. Either way, energy is the key to changing matter from one form to another.

Chemical reactions involve the two basic components of the universe—matter and energy. So far you have seen how matter is involved in chemistry. Now it is time for the energy part.

Energy is the ability to do work. Work is done when matter is moved by a force. Lifting weights is work, as is bending metal or breaking stones. Energy comes in many forms. Electricity, light, heat, and movement are all types of energy. Most

Energy is always being converted from one form to another. Water falls over a cliff and gains motion energy, while lightning converts air movements into electrical energy.

CHEMISTRY IN ACTION
CALORIES

A calorie is a unit of energy. The word *calorie* with a lowercase *c* is the amount of energy needed to heat one gram (0.35 ounces) of water by 1 degree Celsius ($1°C$; $1.8°F$). The word *Calorie* with a capital *C* is used to describe the amount of energy contained in food. One food Calorie is actually equal to 1,000 standard calories. That means a 300-Calorie candy bar contains 300,000 regular calories!

chemical reactions involve adding and releasing heat energy. The study of heat is called thermodynamics. The study of heat energy during chemical reactions is known as thermochemistry.

MOVEMENT AND ENERGY

At the level of individual atoms, there are two main forms of energy involved in chemical reactions: kinetic and potential energy. Kinetic energy is the energy contained by things that move. When you throw a ball, you give it kinetic energy. When you boil a pot of water, you increase the kinetic energy of the water molecules. As the molecules move faster, they break free of each other and become gas. In the same way, when you heat the reactants in a chemical reaction, you increase their kinetic energy and make the reaction proceed more quickly.

Chemists study the kinetic energy of atoms and molecules because it can affect whether or not a chemical reaction will take place. Particles with a lot of kinetic energy move rapidly. When particles are moving quickly, they are more likely to bump into each other. If the right particles hit one another hard enough, they can react with each other and break and make chemical bonds.

POTENTIAL ENERGY

Potential energy is stored energy. A skier at the top of a slope has potential energy. Once he or she begins to ski downhill and move faster and faster, the potential energy is converted into kinetic energy. A battery also has potential energy in a chemical form. When the battery is connected to an electric circuit, the potential energy is converted to electric energy.

A bond between two atoms also contains potential energy. When the bond is broken during a chemical reaction, this energy fuels the reaction and is used to form the bonds of the products. Any energy left over by the reaction is released as heat.

For example, when we eat, our bodies extract useful chemicals, such as sugars, that contain a lot of potential energy. The substances are stored, but when needed they are reacted with oxygen in the blood

A slinky spring moves down a flight of steps. As it is pushed off the top step, the spring's potential energy is converted into kinetic energy and the spring moves to the step below.

PROFILE

JULIUS ROBERT VON MAYER

Julius Robert von Mayer was a doctor who was the first person to realize that living creatures use redox reactions to release energy from food. Von Mayer was born in Heilbronn, Germany, in 1814. After qualifying as a physician, he took a job as a ship's doctor on a voyage to Jakarta, Indonesia. On the long voyage, von Mayer made many observations about heat and temperature. For example, he noticed that rough seas were warmer than calm ones.

In his role as doctor, he noticed that when sailors were injured in warm climates, their blood was brighter red than it was when the weather was cold. Blood carries oxygen from the lungs to the rest of the body, and its color becomes lighter as it picks up more oxygen. Mayer realized that people required less energy to heat their bodies in hotter areas. Their bodies therefore consumed less oxygen, which remained in the blood.

to release their energy. When you are running or exercising in another way, you need to breathe in plenty of oxygen to release the energy that moves your body. That is why you breathe much faster at these times than when you are resting.

TEMPERATURE

The kinetic energy of a group of molecules involved in a chemical reaction is measured as temperature. Temperature is a measure of how fast, on average, the molecules are moving inside a substance. Objects with many fast-moving particles have a high temperature. As the particles slow down, the object becomes cooler.

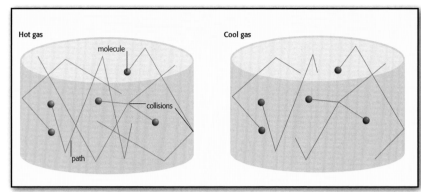

Hot gas
molecule
collisions
path

Cool gas

Gas molecules are always moving. When hot, gas molecules have more kinetic energy and move more quickly. They hit each other and the sides of a container harder and more often than the molecules in a cold gas.

We often think of heat and temperature as the same thing, but chemists see them as two different concepts. When molecules collide with other objects, such as your skin on a hot day, they transfer some of their kinetic energy to the object. That transfer of energy is called heat. On a hot day you are colliding with many moving air molecules and absorbing a lot of their energy. It is this increase in heat that makes you feel warm.

CHANGES IN ENERGY

Energy must be added to break a chemical bond. Just as a boulder at the top of a hill will only begin to roll after being given a push, a chemical bond will not break and release its potential energy without first being given a similar push by added energy. After a chemical reaction has ended, there is almost

TAKING A CLOSER LOOK
CONSERVING ENERGY

Energy cannot be made from nothing or turned into nothing. So, when scientists talk about energy being required or produced during a reaction, they are really talking about energy changing from one form into another. This is the law of conservation of energy, which says that energy cannot be made or destroyed.

For example, a light bulb converts electrical energy into light and heat energy, and when you rub sandpaper on wood it gets hot as the movement energy is converted into heat. Chemical reactions convert energy as well. Burning fuel converts the energy inside chemical bonds into heat.

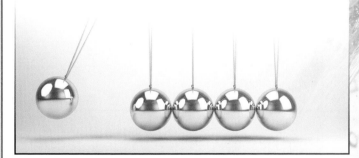

A Newton's cradle shows how the kinetic energy of a moving ball is transferred to a stationary one. One ball stops moving, and the next begins to move.

TOOL AND TECHNIQUES

MEASURING THE ENERGY

To measure the energy stored in a substance, such as fuel or food, chemists use a calorimeter. A sample is sealed in a container and lowered into water. Electricity burns the sample, warming the surrounding water. The rise in the water's temperature is then used to calculate the energy released by the burning sample. Energy is measured in units named joules. One joule (1 J) is the energy used to move 1 kg (2.2 pounds) through 1 m (3.2 feet) in 1 second.

A dancer swings glowsticks to make patterns of light. The glowsticks work by using an exothermic reaction. The reaction releases energy as light rather than heat.

always a difference between the energy added to the reactants and the energy released by the products. This difference is called the enthalpy. The enthalpy of a reaction depends on the compounds involved, and the strength of the bonds that are being broken or built.

In the chemical reaction $AB \rightarrow A + B$, energy is required to break the AB bond. The energy used to break the bond is added by heating molecules of AB. Once AB breaks into A and B, the bond's potential energy is released as heat. However, breaking the bond releases less heat than the amount needed to break it in the first place. Chemists call this an endothermic reaction because it absorbs heat. *Endo* means "inside" and *thermic* means "heat."

Cold packs used to reduce the swelling in injuries use endothermic reactions. Cold packs are full of reactants kept separate by a barrier. Snapping the pack breaks the barrier so the reactants mix and begin an endothermic reaction.

The reaction absorbs heat from the surroundings. This has the effect of making anything touching the cold pack, such as a swollen ankle, colder. The chemical reaction $A + B \rightarrow AB$, also releases energy when the AB bond is built. This extra energy is given off as heat. Chemists call this an exothermic reaction. The term *exo* means "outside." Heat packs that warm skiers' hands use exothermic reactions. When you snap a heat pack you break a barrier, allowing the reactants to mix and begin an exothermic reaction. Heat is given off, warming things near the heat pack. Heat energy always flows from hot objects to colder ones. So the hot pack's heat is transferred to its surroundings.

Heat packs do not make enough heat to burn, but some exothermic reactions get much hotter. Combustion reactions are a type of exothermic reaction that produces so much heat that flames and explosions are produced.

DIAGRAMMING ENERGY

Chemists write chemical equations to show what happens to matter during a reaction. In a similar way, they produce energy diagrams to show what happens to energy during a reaction (see p. 46). An energy diagram is a graph that shows the amount of energy contained by the molecules of reactants as they collide and react, and how much energy is contained in the molecules of product at the end of the reaction.

The horizontal axis of the graph represents time. The farthest left is

KEY TERMS

- **Endothermic reaction:** A reaction that absorbs heat energy.

- **Enthalpy:** The change in energy during a chemical reaction.

- **Exothermic reaction:** A reaction that releases energy.
- **Heat:** The transfer of energy between atoms. Adding heat makes atoms move more quickly.

- **Temperature:** A measure of how fast molecules are moving.

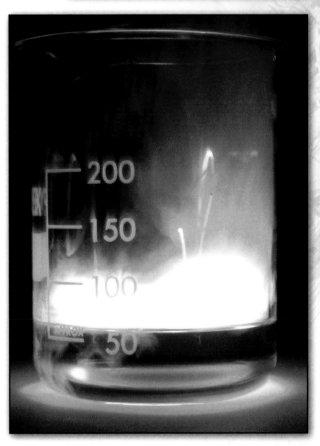

Potassium reacts with water. This reaction is highly exothermic, releasing heat and light in a flash of purple flame.

A burning fuse used to set off explosives. Explosives are chemicals that release huge amounts of energy when they react. The fuse supplies the energy the explosives need to begin reacting.

and the top is high energy. The line that is drawn on the diagram shows the amount of energy at each stage before, during, and after the reaction.

Energy diagrams often look like a hill. The energy starts low, increases and reaches a peak, then reduces again. The top of the "hill" represents the activation energy. That is the amount of energy it takes to make a reaction happen. The activation energy is the minimum amount of kinetic energy needed for the reactants to collide strongly enough for them to react and form the products.

You can read the activation energy from an energy diagram. It is the difference between the energy level of the reactants and the peak.

Energy diagrams also tell you whether a reaction is endothermic or exothermic. If the reactants have higher energy

the start of the chemical reaction, and the farthest right is the finish. As you move from left to right, time moves forward. At first the reactant molecules come together and collide. They then react and form the products. Finally these separate into individual molecules.

The vertical axis represents energy. As you go up this axis, energy increases. The bottom of the graph is low energy

Energy diagrams show how different reactions use energy.

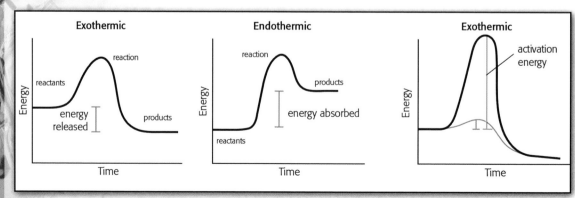

TAKING A CLOSER LOOK
USING THE ENERGY

Cars cannot run without a fuel such as gasoline. Burning gasoline provides the energy needed for the car to move forward, but how does that happen? The engine converts the gasoline's potential energy into movement. It does this in several steps. 1) Gasoline mixed with air is pumped into a cylinder, which is fitted with a piston that can move up and down inside. 2) The piston moves up and squeezes the fuel, making it heat up. 3) An electric spark makes the hot fuel react with oxygen in the air and explode. The reaction produces several gases, which are very hot—the gasoline's potential energy has become heat energy. Hot gases expand very quickly, and the expanding gases push on the piston, forcing it to move down. The heat energy has now been converted to movement, or kinetic energy. The up and down movement of the piston is converted by gears to the spinning motion of the car's wheels. 4) Finally, the piston rises again and pushes out the used gases ready to repeat the process.

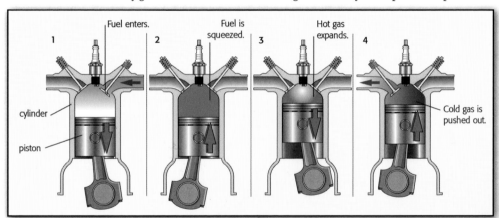

than the products, then overall energy has been released. The reaction is exothermic. If the products have higher energy than the reactants, then energy has been absorbed. The reaction is endothermic.

KEY DEFINITIONS

• **Activation energy:** The minimum energy needed for reactants to change into products.

• **Energy:** The ability to do work.

• **Energy diagram:** Shows the change in energy levels as reactant molecules come together and change into products.

• **Work:** A term scientists use to describe when matter is moved by adding energy.

CHAPTER FIVE

REACTION RATES

The rate of a chemical reaction depends on a number of factors, such as how fast molecules are moving and how tightly packed they are. Chemists control the rate of a reaction by changing these and other factors.

The rate of a chemical reaction is how fast the reactants turn into the products. Some reactions happen very quickly, such as when gunpowder explodes in an instant. Other reactions happen slowly, like a copper statue corroding over many years. The rate of the reaction depends on how the reactants come into contact with each other.

Chemical reactions run at different speeds, or rates, depending on the conditions. Chemists measure how quickly the products are formed from the reactants to calculate the rate.

Reactant particles must physically bump into each other in just the right place for a chemical reaction to take place. In general, small and simple reactants are more likely to hit in just the right place because there is less room for error. Large and complicated reactants have a harder time because they can contact each other in several ways.

To understand this idea better, imagine that you have two sets of shapes: a pair of squares and a pair of octagons. When a particular side of each shape touches the other, the shapes stick together. The chances are it will be faster to find the connecting side of the squares than the octagons. The squares have only four sides to try, while the octagons have eight.

The same thing happens in chemical reactions. Small, simple reactants form products quickly, while large, complex reactants tend to form products slowly. Larger reactants are less likely to collide with each other in the correct position, where they can react.

STARTING A REACTION

Before you can measure the rate of a reaction, the reaction has to get started. Some chemical reactions are spontaneous, meaning the reactions begin on their own. Other reactions need energy to be added before they can start.

To understand how a spontaneous reaction works, think of two rooms connected

TAKING A CLOSER LOOK
RANDOMNESS AND ENERGY

The second law of thermodynamics says that chemical reactions always increase the amount of entropy, or randomness, in a system. This idea can be confusing. The law does not say that the entropy of all the products will be higher than that of the reactants. Some reactions decrease entropy, such as when hydrocarbon gases are used to make plastics. However, the law says that the entropy of the entire universe will always increase. Chemists think big!

The fork and spoon shown here are tarnished with less-ordered silver compounds. Their entropy has increased. People polish their silver to keep it looking bright and new.

Ice melts into water. Inside the ice, the water molecules are arranged in an ordered pattern. As the ice melts, the molecules of water become less ordered and more random. Chemists say that the entropy of the water molecules has increased. If matter is left alone, its entropy tends to increase.

TRY THIS

THE RATE OF RUST

In the rusting activity on page 20, we saw how iron and oxygen reacted to make iron oxide, or rust. You can repeat this reaction to investigate how different conditions affect the rate of reaction. You'll need three jars and three new nails. In jar 1 repeat the activity as before. Ask an adult to cover the nail in jar 2 with boiled water and put the lid on. In jar 3, place a nail with no water and put the lid on. Return after a day and compare the results.

By changing the conditions, you change the rate of the reaction. The nail in jar 1 will have rusted as before. The boiled water in jar 2 contains very little oxygen, so hardly any rust will form. The reaction runs more quickly in wet conditions, so very little rust will have formed in jar 3. However over a long time, moisture in the air may help the nail to rust.

by a closed door. One room contains air fresheners to make the room smell like flowers, the second room does not. When you open the door between the two rooms, some of the air freshener will naturally flow into the second room. Soon both rooms will smell like flowers, but the scent will not be as strong. Nothing is pushing or blowing the scent, it just spreads out. This movement is caused by what chemists call entropy.

Entropy is a measure of how random, or disordered, a system is. Most things in nature lack order, including many things in chemistry. The oxygen atoms in the air are not stacked neatly. They move randomly. Disordered systems have high

CHEMISTRY IN ACTION
BACTERIA REACTIONS

When you put your leftover dinner in the refrigerator, you slow the chemical reactions that threaten to spoil your food. We say a food is spoiled, or rotten, when unwanted bacteria grow in the food, making it taste bad. Spoiled food can also make you ill.

The bacteria use chemical reactions to grow. Putting the food in a refrigerator lowers the food's temperature, slows the rate of the reactions, and slows the growth of bacteria. This means your food takes longer to spoil, and you can eat the leftovers later.

entropy. How much entropy something has depends on its state of matter and on the surrounding conditions. With a lot of atoms or molecules moving around, a gas has high entropy. With many atoms packed in place, a solid has much less entropy. The entropy of a liquid is somewhere in between.

In a chemical reaction, the entropy of the reactants partly determines how the reaction begins and how fast it goes. A spontaneous reaction is driven by entropy. Just as the air freshener moves from one room to another, reactants move together and form products in a spontaneous reaction.

Reactions that are not spontaneous require something to get started. Usually, energy is added to the system to make the reactant molecules move fast enough so that they can react when they collide.

How fast a chemical reaction forms products depends on how often and how hard the reactants bump into each other. There are two things that determine the number of collisions between reactants: concentration and temperature.

THE CONCENTRATION OF CHEMICALS

Imagine the octagons with sticky sides again. If you have only two octagons moving about randomly in a room, like gas particles, it may take those octagons a long time to bump into each other in such a large space. However, if you increased the concentration—the number of octagons in the same space—you would have better luck. A thousand octagons in the same room will be much closer together and therefore hit each other far more often. Because there are more collisions between octagons, pairs will form more quickly.

In most cases the concentration of reactants affects how fast the reaction will occur because it determines how likely reactants are to bump into each other in the right way. If the concentration of reactants is high, the reaction will

Chilling food makes all chemical reactions inside the food stop almost completely. Frozen food can be kept for several weeks and still be good to eat.

be fast. If the concentration of reactants is low, the reaction will be slow.

If a reaction involves several reactants and takes place in a series of steps, this general rule about the effect of concentration may no longer be true. The reaction can only proceed as quickly as its slowest stage. Therefore it is the concentration of just the reactants involved in that stage that determines the rate.

Increasing the concentrations of the other reactants will have no effect.

HEAT AND REACTANTS

Just as the concentration of the reactants affects the chance of colliding in the right spot, so does the temperature. Remember that temperature is a measure of how fast atoms or molecules move. Cold temperatures show that molecules are moving slowly. High temperatures occur when molecules move quickly.

Cold, slow-moving molecules are less likely to collide and begin the chemical

TAKING A CLOSER LOOK
THE SIZE AND SPEED OF A REACTION

When one of the reactants is a solid, the size of its pieces will affect the speed of the reaction. The reaction takes place on the surface of the solid, where the reactants come into contact. A large solid has a small surface area compared with its weight. Most of the reactant is locked away inside the solid and cannot take part in the reaction. Therefore the reaction is slow. If the solid is crushed into a fine powder, its total surface area is hugely increased. More of the reactants can take part in the reaction, which now takes place much more quickly.

Calcium carbonate reacts with acid, producing carbon dioxide gas. The beaker on the left contains powder, while the one on the right contains a single piece of the carbonate. Carbon dioxide is being produced more quickly in the left beaker, making it fizz.

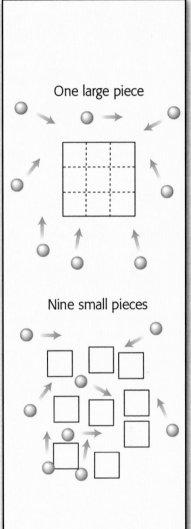

One large piece

Nine small pieces

A diagram of how the same amount of solid will come into contact with other reactants more quickly when broken into pieces.

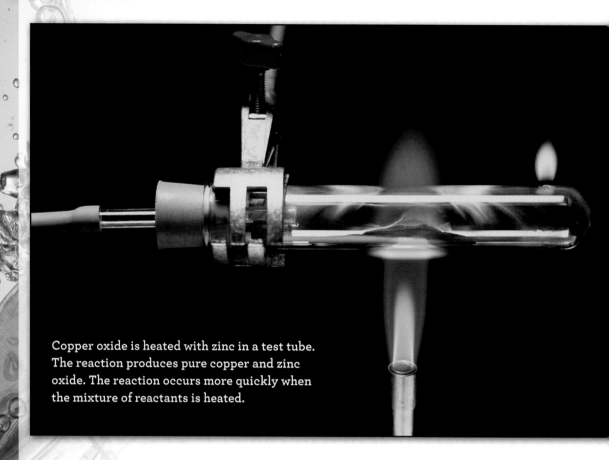

Copper oxide is heated with zinc in a test tube. The reaction produces pure copper and zinc oxide. The reaction occurs more quickly when the mixture of reactants is heated.

reaction. That is why food kept cold will not react with the air and spoil. Hot, fast-moving atoms are more likely to collide in the right way and react. This is also why food cooks faster at higher temperatures (but too high and it burns). However, that is not the whole story.

The energy that atoms possess at different temperatures also affects reaction rates. At low temperatures, atoms often do not have enough energy to form bonds, even if they do collide in the right place. At high temperatures, atoms are not only more likely to collide, they are also more likely to

have enough energy to get the reaction going when they do.

HOW DO YOU CONTROL A REACTION?

Concentration and temperature are two things that affect the rate of a reaction. These factors can be altered during a reaction, and they play a part in all reactions. Chemists can also introduce substances to slow or speed up the rate of reaction. A catalyst is something that speeds up reactions. An inhibitor is something that slows or stops reactions. These are neither reactants nor products, and are unchanged by the chemical reaction taking place. Catalysts and inhibitors work by changing the activation energy—the amount of energy the reactants need to begin the reaction. However, they do not change how much energy is released or absorbed by a reaction.

Catalysts decrease the activation energy, so more of the reactants form products at low temperatures. Catalysts often work by providing a surface where reactants come together and react without requiring a lot of energy. For example, the catalytic converters in cars use a thin layer of platinum and iridium as catalysts. These metals help remove the harmful gases produced by burning gasoline and diesel fuel. The harmful gas molecules cling to the surface of the catalysts, where they can react with oxygen and become less dangerous gases. For example, the converter changes carbon monoxide gas (CO), a poison, into carbon dioxide (CO_2).

Inhibitors increase the activation energy. As a result fewer of the molecules collide with enough energy to form products, and the reaction slows. Inhibitors often work by bonding with the reactants, so other molecules cannot make contact. Many reactions that take place inside living cells are controlled by inhibitors.

CHAPTER SIX

WHAT IS ELECTROCHEMISTRY?

Chemical reactions can generate electricity. Reacting electrons are forced to flow in one direction, creating an electric current. The current is then used to power machines.

Electrochemistry uses redox reactions to produce electricity. Remember that a redox reaction happens when electrons move from one compound to another. In electrochemistry, the electrons that are exchanged during redox reactions are forced to travel from one place to another. These traveling, charged particles create an electric current.

Redox is short for reduction-oxidation, the two halves of the reaction. The reduction occurs when one compound gains electrons. The oxidation occurs when one compound loses electrons. Chemists use something called a voltaic cell to harness the electricity produced by the reactions. The voltaic cell is named for Alessandro Volta (1745–1827), an Italian count who invented

Batteries contain chemicals that produce electric currents when they react.

the device in 1799. The volt (V), the unit used for measuring the strength of an electric current, is also named for him.

WHAT IS A VOLTAIC CELL?

A voltaic cell is a pump that forces electrons to move from one place to another. In a simple cell, there are two dishes of chemical solutions connected by a U-shaped pipeline. One dish is where the reduction reaction takes place; the other dish is where the oxidation reaction takes place. The electrons must travel between the two dishes through the pipeline. The pipeline is filled with a solution of ions to carry the electrons.

Each dish contains a stick of metal called an electrode. Metals are good conductors of electricity because they have many free-floating electrons inside. The metal electrodes serve as a holding tank of electrons, ready to carry a current when needed once they are connected to each other with a wire.

The electrode where the reduction reaction takes place is called the cathode. The cathode is usually labeled with a "–" because the current of electrons flows away from it. The oxidation-reaction electrode is called the anode. The anode is labeled with a "+" because the current flows toward it.

The electrodes are immersed in a liquid that contains the reactants for

PROFILE
MICHAEL FARADAY

English chemist Michael Faraday (1791–1867) is remembered for investigating electrochemistry. The son of a blacksmith, Faraday taught himself about chemistry by reading books and by helping out in laboratories as a boy. As an adult he established the laws of electrolysis and established the link between electricity and magnetism. He made the first electric motor and generator and made up the terms *electrolyte, electrode, anode,* and *cathode.*

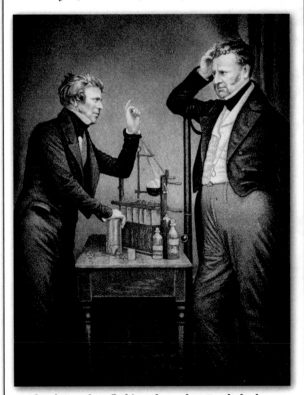

Michael Faraday *(left)* and another English chemist, John Daniell, in Faraday's laboratory. Daniell is remembered for inventing a type of battery.

the redox reaction. The two halves of the reaction happen continuously until one of the reactants runs out. In the reduction reaction, a reactant gains electrons from the anode to become the product. This product then joins onto the anode, gradually building up a thin layer. In this way the anode is constantly supplying the reactants with electrons. At the cathode the opposite happens. The reactant is being oxidized and is giving away electrons. The cathode collects these electrons.

The electrons flow through the wire between the electrodes. Electrons are negatively charged, so they flow away from the negative cathode to the positive anode. The current contains energy and this energy can be transported along wires to provide power to run machines.

Voltaic cells can use many different types of chemicals in the redox reactions, but they often contain metals. One reaction involves zinc and manganese dioxide. The pure zinc is oxidized into zinc hydroxide and releases electrons.

The manganese dioxide is reduced into manganese trioxide and gains electrons.

BATTERIES TODAY

A battery is a portable voltaic cell. Batteries come in many shapes, sizes, and powers, but they all work by storing chemical energy and turning it into electrical energy using a redox reaction.

All batteries contain an anode (the flat, negative end) and a cathode (the raised, positive end). Instead of a U-shaped tube, the electrodes are separated by a mixture of chemicals called electrolytes. Electrolytes contain ions. Ions are formed when atoms gain or lose electrons. This gives the atoms a charge. Positive ions have lost electrons, while negative ions have gained them. The ions flow through the electrolyte and carry the charge between the electrodes.

When a battery is sitting on a table, the redox reaction is not happening inside because there is nothing

TOOL AND TECHNIQUES
AMMETERS

An ammeter (*right*) measures the size of an electric current. It contains a coil of wire that is surrounded by a magnet. When an electric current passes through it, the coil produces a magnetic force that pushes against the magnet. This force makes the coil twist around and move a needle. The needle moves along a scale that shows how large the current is. When the current is switched off, the coil and needle are pulled back into the starting position again by a spring. Electric current is measured in amperes (A). One ampere is a current where 6.25 million trillion electrons pass through a circuit every second.

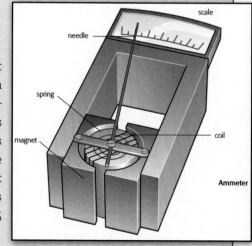

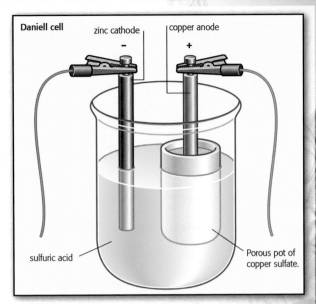

Daniell cell — zinc cathode — copper anode — sulfuric acid — Porous pot of copper sulfate.

The Daniell cell is a type of voltaic invented by John Daniell (1790–1845). The anode is solid copper and the cathode is zinc metal. The cell has two electrolytes. The anode is in copper sulfate solution, while the cathode is in sulfuric acid. The two liquids are separated by a porous barrier. The zinc atoms in the cathode are oxidized. They lose two electrons and form zinc ions (Zn^{2+}). These ions dissolve in the acid. The copper ions (Cu^{2+}) in the copper sulfate are reduced. They pick up electrons from the anode and form copper atoms, which join on to the anode.

The electrons released in the cathode move through a wire to the anode, producing an electric current of 1 volt.

connecting the anode and cathode. When you put your battery into a device, such as a flashlight, you create a path between the anode and cathode. That completes a circuit and the redox reaction begins. The circuit is a circular path through which electricity can flow. If there is a break in the circuit the electricity will not flow.

When the redox reaction begins inside the battery, electricity is produced. A fixed amount of reactants is stored in the battery. When the reactants are completely used up, we say the battery is "dead." A dead battery can no longer produce electricity because the chemical reaction is not taking place.

Some batteries are rechargeable. They work in the same way as a regular battery. However, when a rechargeable

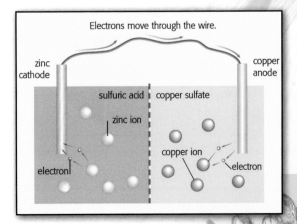

Electrons move through the wire. zinc cathode — copper anode — sulfuric acid — copper sulfate — zinc ion — copper ion — electron — electron

Modern batteries are more complicated than voltaic cells but they use electrochemical reactions in the same way. In this example, the cathode is a paste containing manganese dioxide. This paste is also the electrolyte. The anode is powdered zinc. A metal rod transfers electrons from the cathode to the rest of the circuit.

cathode   anode

metal rod

battery becomes dead, you plug it into an electrical outlet. Electricity from the outlet forces all the reactions inside a rechargeable battery to run in reverse. The anode becomes the cathode, the cathode becomes the anode, and the products of the reaction become the reactants. This generates a new source of the original reactants so the battery can make electricity again.

WHAT ARE ELECTROLYTIC CELLS?

Electrolytic cells are the opposite of voltaic cells. Voltaic cells use a redox reaction to produce electricity. Electrolytic cells use electricity to produce a redox reaction. Electrolysis is a process of pushing electricity through a solution to force a reaction to occur. The reaction generally involves a stable compound being broken apart. The word *electrolysis* means "splitting with electricity." During the electrolysis of water, for example, water molecules are broken into their hydrogen and oxygen atoms. Chemists

TRY THIS
MAKING A PILE

Electric batteries were once called piles because they were made from stacks, or piles, of voltaic cells. You can make your own pile using a stack of pennies and dimes, paper towel, and some concentrated lemon juice. The coins are the electrodes; the penny is the anode and the dime the cathode. The lemon juice is the electrolyte.

Cut the paper towel into about ten 1-inch (2.5 cm) squares. Soak these pieces in the lemon juice so they are sopping wet. Use a strong paper towel so it does not break apart when wet. Clip one end of an electric wire to a penny, and begin building a stack in this order: penny, dime, paper towel, penny, dime, paper towel, and so on. Use all the paper towel pieces and end with a dime on top. Connect a second wire to this top dime.

Connect the loose ends of the wires to an ammeter. Switch the wires if you do not get a reading. You can repeat the experiment using salty water instead of lemon juice. The piles could also be made with aluminum foil and iron nails. These changes will result in a slightly different size of electric current.

An electric battery made from pennies and dimes. A pair of coins on top of each other makes a single cell. The paper towel is used to divide one cell from another. The juice can pass through the paper, and so an electric current will run through the whole pile. Connect an inexpensive ammeter or voltmeter to see how much current is produced.

Water is split into hydrogen and oxygen gas using electrolysis. Gas bubbles form on electrodes and are collected in upside-down test tubes. Twice as much hydrogen *(left)* is produced as oxygen.

molecules to make oxygen atoms and hydrogen ions (H^+):

$$2H_2O \rightarrow O_2 + 4H^+ + 4 \text{ electrons}$$

The oxygen bubbles out of the liquid like the hydrogen. The electrons travel along a wire to the cathode and take part in the reduction of more water molecules.

WHAT IS A FUEL CELL?

A hydrogen fuel cell uses hydrogen and oxygen gas to make electricity. A hydrogen fuel cell is like a battery, but has a different source of reactants. A battery has reactants stored inside. Space is limited, and when the reactants run out, the battery becomes dead. The reactants for a fuel cell are stored outside and pumped into the cell, so space is not a problem. In a hydrogen fuel cell, hydrogen and oxygen are combined to form water. As they react, the energy released is used to produce electricity. Fuel cells are an attractive source of power because they do not produce any harmful pollution, just water vapor.

USING ELECTROLYSIS

Electrolysis is used to separate pure metals from their compounds. That is called

run electricity through liquid water to start the reaction, which has the basic equation:

$$2H_2O \rightarrow 2H_2 + O_2$$

Like all redox reactions, the electrolysis of water has two halves. The water is reduced at the cathode. The molecules receive electrons. This causes them to split into hydrogen gas (H_2) and hydroxide ions (OH^-):

$$2H_2O + 2 \text{ electrons} \rightarrow H_2 + 2OH-$$

The ions stay in the the water, but the hydrogen forms gas bubbles on the electrode that float up out of the water.

At the anode the water is oxidized: Electrons are released by the water electrorefining. The process used is the same as the one to break water apart. In nature, most metals are found as compounds. Electrolysis is often the only way to extract reactive metals.

The metal compounds are dissolved or melted to make an electrolyte. Electrolysis provides the energy needed to remove the metal atoms from the compound. The metal atoms collect at one electrode. Waste material forms at the other.

The same electrolysis process is used to cover objects with a thin layer of metal. This is called electroplating. When you buy an inexpensive gold necklace, it is likely to be another type of metal covered with a very thin layer of gold.

During the electroplating process, the object to be coated becomes the cathode. The valuable metal used for coating is the anode and is also dissolved in the

A necklace that has been plated with gold. Inexpensive items of jewelry are coated with a very thin layer of precious metal by electroplating.

electrolyte. Electrolysis deposits the anode metal onto the cathode.

CHEMISTRY IN ACTION
BATTERIES AT HOME

The batteries that you put in flashlights, radios, remote controls, and other portable electronic devices contain several voltaic cells stacked on top of each other. The stacked cells, called a pile, can generate more electricity than a single cell. The size of the battery indicates something about the number of cells inside. A 9-volt battery has six 1.5-volt cells in a pile. A 12-volt car battery has six 2-volt cells in a pile.

This 9 volt battery is made up of six cells piled on top of each other. Each cell is 1.5 volts.

CHAPTER SEVEN

WHAT IS A NUCLEAR REACTION?

Nuclear reactions turn one element into another element, releasing enormous amounts of energy in the process. Chemists study how to control nuclear reactions safely so that they can be used to create energy.

Nuclear reactions are reactions that involve the particles inside the nucleus of an atom. All chemical reactions involve only the electrons in the space surrounding the nucleus. None affect the nucleus itself. Nuclear reactions are different because they change

The Sun's heat and light are produced by nuclear reactions at its center.

one element into another by changing the number of protons in the nucleus of the atom.

INSTABILITY

Nuclear reactions involve a set of elements that chemists describe as radioactive. Radioactive elements have unstable atoms. The nuclei of these atoms often break apart, especially when they collide with another particle, such as a neutron. Elements with 83 or more protons in the nucleus are the most radioactive.

The instability is a result of there being so many protons packed into the nucleus. Protons are positively charged, so they are constantly pushing against each other. The reason they do not fly apart from each other is a stronger force that holds the protons and neutrons together. This force, called the strong nuclear force, only works over tiny distances. It has no effect outside the nucleus. In the nucleus of an unstable atom, the strong force is not powerful enough to keep all the particles together. Eventually the nucleus begins to break apart, or decay.

Radioactive decay occurs when the nucleus gives off small particles. These particles are also often called radiation. There are three types of radiation, each named after Greek letters: alpha, beta, and gamma. An alpha particle is two protons and two neutrons, a beta particle is a single electron, and a gamma ray is an emission of energy.

CHEMISTRY IN ACTION

WHAT IS NUCLEAR POWER?

Nuclear power plants use the energy released by nuclear reactions to generate electricity. The heat from the reactions is used to boil water and make steam. The steam is used to spin large fans called a turbine. The turbine's spinning motion is converted into electricity by a generator. Coal and gas power plants do the same thing; they just use coal or oil fires to heat the water.

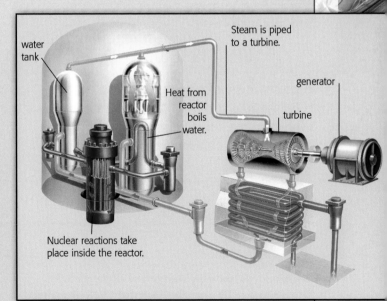

water tank

Steam is piped to a turbine.

Heat from reactor boils water.

generator

turbine

Nuclear reactions take place inside the reactor.

A diagram showing how the heat from nuclear reactions is used to make electricity.

A power plant worker wears a suit, gloves, and a hood to protect himself from radiation leaking from a nuclear reactor. He breathes through a mask that filters radioactive particles from the air.

WHAT IS RADIATION?

If you have already heard of nuclear reactions, it is probably because of the health concerns over radiation. High doses of radiation damage living cells so they die or cause dangerous illnesses.

Nuclear radiation burns the skin in the same way sunlight causes sunburn. Inside the body, radiation is more dangerous. It breaks up the contents of cells so they no longer

A photograph showing the path of alpha particles being released by a speck of radium. The radiation is released equally in all directions.

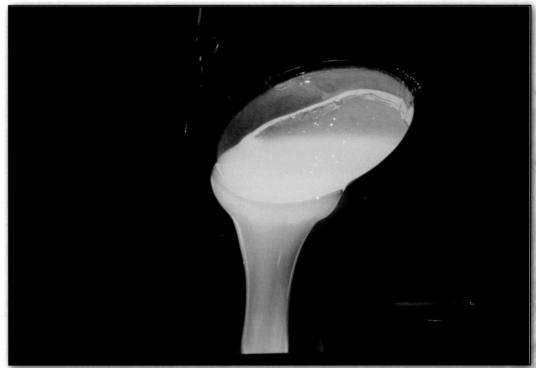

Melted glass containing radioactive waste from a nuclear reactor is poured into a mold. The glass prevents radiation from leaking out. This technique is being investigated to see if glass is a safe way to store the dangerous radioactive waste produced by nuclear reactors.

work properly. The radiation may damage the cell's DNA, which causes the cell to work in the wrong way. DNA stands for deoxyribonucleic acid, which is the molecule that contains the information needed to organize a living cell. Damaged DNA causes illnesses. For example the cell might begin to grow too quickly and become a cancerous tumor.

Radiation spills are a huge concern when building nuclear power plants. If something goes wrong in a nuclear power plant, radiation can be released into the air and harm living things in the area. For example, in 1986 an explosion at a reactor in Chernobyl, Ukraine, spread radiation over the homes of 200,000 people. All of them had to be evacuated.

Another major concern with nuclear power plants is the radioactive waste they produce. Nuclear reactions continue in this radioactive waste and can continue to emit dangerous amounts of radiation for 10,000 years or more. The waste

TOOL AND TECHNIQUES
GEIGER COUNTERS

A Geiger counter is used to measure all types of radiation. Inside the Geiger counter is a tube filled with gas. When radiation enters the tube, it knocks electrons off the gas molecules, turning them into ions. The electrons are picked up by a wire running inside the tube and produce a pulse of electricity. Each pulse shows that one particle of radiation has collided with the gas in the tube.

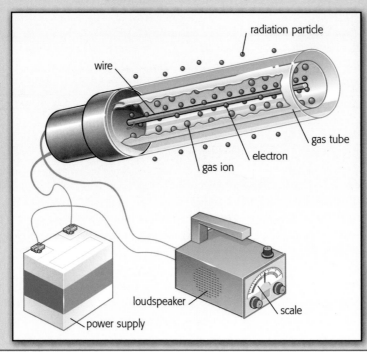

radiation particle

wire

gas tube

electron

gas ion

loudspeaker

scale

power supply

This diagram shows how a Geiger counter works. The counter shows how much radiation is in the tube on a scale and by producing clicks.

must be stored for all that time. Despite the concerns over safety and the costs of storing waste, nuclear power produces less pollution than other methods.

WHAT ARE ISOTOPES?

Certain forms of the elements are more radioactive than others. The different forms of atoms are called isotopes. The number of protons in an atom determines what element it is. However, some atoms of the same element have different numbers of neutrons. This makes them different isotopes. Some isotopes are more radioactive than others because the number of neutrons makes them unstable.

Physicist Guy Savard examines a gas catcher cell that is part of a centrifuge that generates intense beams of nuclear isotopes.

To learn more about isotopes, it helps to understand how they are written. Chemists write isotopes in two ways. One way to write an isotope looks like this: $_z^a X$, where "X" is the chemical symbol, "z" is the atomic number (the number of protons in the nucleus), and "a" is the atomic mass number (the sum of the protons and neutrons in the nucleus). If you subtract the atomic number from the mass number, you get the number of neutrons in that isotope. This level of detail can be useful but it is not always needed. Chemists may just write the symbol and the mass number, such as U-238 (uranium-238, or $_{92}^{238}U$).

When some isotopes start to decay, they sometimes cannot stop. For example, U-238 is an unstable isotope of uranium. When it begins to decay, it produces thorium-234 (Th-234), another unstable isotope. Th-234 decays to protactinium-234 (Pa-234), yet another unstable isotope. The decay continues creating unstable isotopes until, after 14 unstable isotope steps, a stable atom is formed and the decay stops. Such a series of steps is known as a radioactive decay series and is not unusual among the most unstable elements.

WHAT ARE FISSION REACTIONS?

One type of nuclear reaction is called a fission reaction. A fission reaction

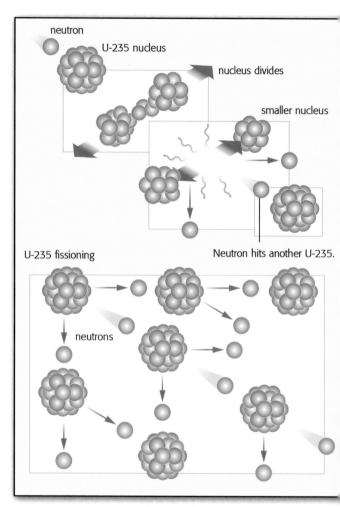

occurs when a neutron strikes a large nucleus, breaking it into two or more new elements with smaller nuclei. In general, the equation for a fission reaction looks like this:

$$_{z}^{a}W + n \rightarrow \, _{z}^{a}X + \, _{z}^{a}Y + n$$

Element W is bombarded by a neutron (n), and produces two new elements, X and Y, plus more neutrons. As before, the numbers "a" and "z" represent the atomic mass and atomic number. (The atomic mass of a neutron is 1 and its atomic number is 0.)

Unlike a chemical reaction, which creates bonds between different elements, a fission reaction actually creates new elements. Fission reactions also release a lot of heat and several neutrons. These neutrons are then free to bombard other radioactive nuclei and start more fission reactions. That is a chain reaction. The nuclear reactors at power plants control the chain reaction so that the heat and radiation are released slowly and safely.

A diagram of how a nuclear fission can cause a chain reaction. A neutron hits a uranium-235 (U-235) nucleus. The nucleus splits in two and releases two or three neutrons. The neutrons then hit more U-235 atoms and cause yet more fission reactions.

WHAT ARE FUSION REACTIONS?

A fusion reaction is the opposite of a fission reaction. A fusion reaction combines two small nuclei to create a single large nucleus and release a lot of energy. Fusion reactions are very hard to start because they take huge amounts of energy to begin.

One place where fusion reactions happen continuously is inside the Sun. The Sun turns hydrogen atoms into helium by fusion, and releases the energy that lights and warms our planet.

On Earth, scientists have experimented with ways to generate power from fusion reactions. Because the reactions require so much energy to get started, they are not easy to investigate. Fusion reactors are being built to test whether it is possible to release more energy from a fusion reaction than is added to start it. If this is achieved, fusion could be a useful new source of power.

NUCLEAR ENERGY

You already know that the two fundamental components of the universe, matter and energy, interact during a chemical reaction. In a nuclear reaction, both matter and energy are still present and necessary, but the rules are different. Matter can be changed from one element to another during a nuclear reaction. But matter can also be changed into energy.

So much energy is released by a nuclear reaction because some of the matter in the original nucleus is converted into energy. Because chemists know the amount of mass involved, they can calculate the energy using the famous equation, $E = mc^2$. E is energy, m

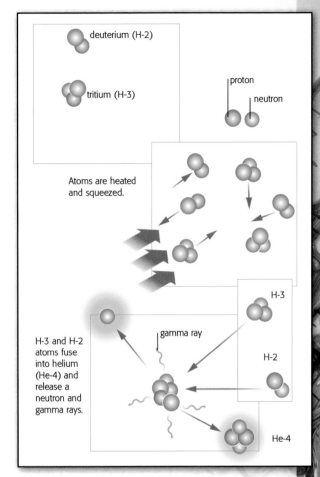

A diagram of the fusion reaction that takes place in the Sun. Two radioactive hydrogen isotopes—deuterium and tritium—fuse to make a single helium atom. Normal hydrogen atoms (H) have a single proton as a nucleus. Deuterium (H-2) has one proton and one neutron in the nucleus, while tritium (H-3) has one proton and two neutrons.

is mass, and c is the speed of light. This equation was figured out by the great German-born American scientist Albert Einstein (1879–1955). The speed of light, it turns out, is a huge number and as a

KEY DEFINITIONS

• **Atomic number:** The number of protons in the nucleus.

• **Fission:** When a large atom breaks up into two small atoms.

• **Fusion:** When small atoms fuse to make a single larger atom.

• **Isotope:** Atoms of the same element with different numbers of neutrons. Many isotopes are radioactive.

• **Radioactive decay:** When small particles break off from an unstable nucleus.

result small amounts of mass produce huge amounts of energy.

NUCLEAR EQUATIONS

To balance a chemical equation you count atoms. To balance a nuclear equation you count the the number of subatomic particles.

The sum of all the atomic particles' numbers on the left-hand side must

Albert Einstein (*left*) in discussion with U.S. physicist Robert Oppenheimer. Oppenheimer was head of the Manhattan Project, which built the first nuclear bomb. The project used Einstein's theories about matter and energy to make the bomb.

TAKING A CLOSER LOOK
ATOM BOMBS

The first atomic bomb was dropped by the United States on Japan in 1945. It contained two blocks of uranium metal that were kept separate. The bomb was set off by a normal explosive, which shot the smaller metal block onto the larger one. Together, the blocks made a critical mass—a lump of uranium large enough for fission reactions to start. An uncontrolled chain reaction of fissions released heat and radiation in a huge blast. The bomb killed 140,000 people. Some people died in the explosion, but many more died afterward from the effects of radiation.

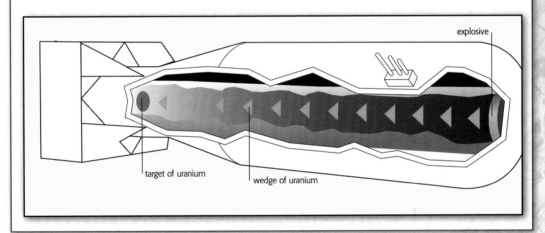

explosive

target of uranium

wedge of uranium

equal the sum of all particles on the right-hand side. None of the elements bond to make compounds, instead one element is changed into another element.

For example, when a uranium-235 (U-235) atom is hit by a neutron (n), a fission reaction occurs. U-235 breaks into two smaller elements, barium (Ba-142) and krypton (Kr-91):

$$^{235}_{92}\text{U} + \text{n} \rightarrow {}^{142}_{56}\text{Ba} + {}^{91}_{36}\text{Kr} + 3\text{n}$$

This equation is balanced. The number of particles in the reactants (235 + 1) is equal to the products (142 + 91 + 3).

CHAPTER EIGHT

BIOGRAPHY: FRITZ HABER

Fritz Haber was a german physical chemist. His name has passed into the language of chemistry by way of the "Haber process," a method of manufacturing the gas ammonia. Ammonia was used to make artificial fertilizers and explosives—substances that Germany was desperately short of during World War I.

Fritz Haber was born to a German-Jewish family in Breslau in Silesia in 1868. Silesia was then part of Prussia. Three years later, Prussia became the leading power in the newly formed German Empire, which was known as the Second Reich.

Haber's father was a successful merchant trading in natural dyes, and a major importer of the natural blue dye indigo, which was obtained from plants that grew in Asia. After studying at the German universities of Berlin and Heidelberg, and at Zurich in Switzerland, Haber joined the family business as a traveling salesman.

Haber soon realized that dramatic changes were taking place in the dye industry. The future, he was sure, lay with synthetic dyes (dyes that are produced artificially through a chemical reaction). The market for the plant madder, a source of red dye, had collapsed with the development of the synthetic red dye alizarin in 1868, and Haber realized that it would not be long before someone manufactured a synthetic replacement for indigo.

Because of disagreements with his father, Haber left the company and got a junior post in the Department of Chemical and Fuel Technology at the Karlsruhe Technical University, in southwest Germany. Haber's work at Karlsruhe would have two dramatically different effects. On the one hand it would save the world from starvation. On the other, it would create a way of killing and disabling many thousands of people quickly and efficiently.

FOOD AND WEAPON PRODUCTION

In his book *Chemistry in its Applications to Agriculture and Physiology* (1840), German chemist Justus von Liebig (1803–1873) concluded that among the "foods" needed by plants for healthy growth were nitrogen and potassium. He mistakenly thought that plants could get nitrogen from the air. In fact, plants absorb nutrients from the soil. When land is intensively farmed year after year, unless nutrients are returned to the soil through fertilizers, plant yields get lower and lower.

Traditional methods of farming used animal manure as a fertilizer, but the world's population was rapidly growing at the end of the 19th century, demanding an ever-increasing food supply. More and more land was coming into production, so huge amounts of fertilizer were needed. The question was, where could supplies be found?

For a time the vast deposits of Chile saltpeter (sodium nitrate) found in the Atacama region of northern Chile provided the answer. By 1913 Chile was responsible for 90 percent of the world's supply of nitrogen-based fertilizers, with

Key Dates	
1868	December 8, born in Breslau, Prussia
1901	Marries Clara Immerwahr, the first woman to gain a PhD in science at Breslau University
1908	Appointed professor of chemistry at Karlsruhe Technical University
1911	Becomes director of the Kaiser Wilhelm Institute for Chemistry, Berlin
1916	Appointed director of the chemical warfare service
1918	Awarded Nobel Prize for Chemistry
1933	Is forced to leave Germany
1934	January 29, dies in Basel, Switzerland

Germany taking about one-third of the annual production. But these deposits, vast though they had once been, were running out. At the start of the 20th-century scientists had begun to forecast doom, warning that a chemical solution must be found if people were to be properly fed.

The problem was particularly serious in Germany. In the years leading up to World War I (1914–18), it was concentrating its economic resources on building its military capabilities and becoming a leading world power. Nitrates (nitrogen-containing salts) were not only essential ingredients in fertilizers, they were also a vital component in many explosives. Leading figures in Germany soon realized

Berthold Schwarz was said to have introduced gunpowder. His surname means "black" and was a nickname he presumably earned from carrying out explosive chemical experiments.

GUNPOWDER: THE ORIGINS

The first significant explosive, gunpowder, was known in China from about the 9th century AD and in the West from the 13th century. The English philosopher and scientist Roger Bacon (c. 1214–1292) wrote down a coded recipe for gunpowder, which, translated into modern terms, indicates that he used a mixture of charcoal (carbon), sulfur, and saltpeter (potassium nitrate). The earliest illustration of gunpowder being used in Europe dates from 1327, and shows a curious arrow-shaped missile being fired from a cannon.

The development of gunpowder led to major changes in warfare. Stone castles were no longer safe against the powerful guns of a besieging army. Even the mighty walls of Constantinople were defenseless against the cannons of the Turkish army in 1453. From the 15th century, a number of so-called "gunpowder empires" developed as European powers, backed up by cannon carried on board their ships, seized territory in Asia, Africa, or America.

A KEY INGREDIENT

The key ingredient in making gunpowder was saltpeter, a nitrate. At first nitrates were extracted from decomposing animal and vegetable matter—mainly urine and manure. The supply of nitrates was so important that in most European countries it was controlled by the state, and saltpeter workers were given key privileges. In 1625 in England the law allowed saltpeter workers to enter any property and remove any material likely to contain nitrates. In France after 1540 saltpeter workers could claim free lodgings and transport. But supplies remained low, and the lack of saltpeter for making gunpowder was one of the factors that brought an end to the Seven Years' War in Europe in 1763.

A more convenient and abundant supply turned up at the start of the 19th century with the discovery of vast beds of Chile saltpeter (sodium nitrate) in northern Chile. The deposits were found by a local woodcutter, who built a fire and found that the ground below it started to melt. He reported the strange event to the authorities, who realized what it must be. Mining started in about 1825. Over a century later, when the industry was well in decline, Chile was still producing more than 1 million tons a year. Once Haber had discovered how to manufacture nitrates artificially, however, there was no longer a problem of supply.

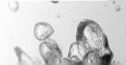

that, in the event of a major war, German ports would be blockaded to cut off fresh supplies of nitrates. With no extensive nitrate supplies of its own, Germany would be unable to grow sufficient crops to feed its people properly, or to supply its army with enough explosive ammunition to fight its enemies.

Nitrates could be recovered from ammonia, which is a compound of nitrogen and hydrogen. Supplies of ammonia could be obtained from coal, but only 5 pounds (2.27 kg) of ammonia could be taken from about one metric tonne (1,000 kg) of coal, making it very expensive. It was obvious that other sources would have to be found.

CREATING AMMONIA

What was needed was a way of making ammonia artificially—in other words, synthesizing it from nitrogen in the atmosphere and hydrogen extracted from coal gas. Earlier attempts to make nitrogen react with hydrogen using an electric spark failed when it was found that a temperature of up to 3,000°C (5,432°F) was needed.

Experts warn that nitrate fertilizer, used to help crops grow, leaches into the groundwater and can cause water pollution 80 years after it is initially used.

When chemists find it difficult to produce a desired chemical reaction, they often resort to the use of catalysts. A catalyst is a substance whose presence in small quantities can speed up a chemical reaction. The catalyst itself is not used up during the reaction. For example, poisonous carbon monoxide and oxygen will hardly react at all at room temperature, but with the aid of a sprinkling of platinum powder as a catalyst the gases readily combine to form carbon dioxide. This is what happens in the catalytic converter that is a part in many automobiles.

Haber therefore tried to combine nitrogen and hydrogen by finding a suitable catalyst. At first he thought that the rare metal osmium was the ideal catalyst, and he acquired exclusive rights to purchase the world's entire stock of 220 pounds (100 kg). But after testing 4,000 other catalysts, he found that—suitably treated—iron, much more abundant and therefore cheaper, would work just as well. Consequently in 1903 Haber succeeded

Nitrates were essential for making explosives

in combining nitrogen and hydrogen at a temperature of 1,000°C (1,832°F) using an iron catalyst. The yields, however, were too small to be commercially useful. A colleague, Walther Nernst (1864–1941), pointed out that increasing the pressure might help the reaction, and after further calculation Haber worked out how much extra pressure would be needed to get a productive reaction.

By 1909 Haber had established that he could produce large quantities of ammonia with an iron catalyst at a temperature of 932°F (500°C) and a pressure of about 200 atmospheres. An atmosphere is a unit of pressure equivalent to 14.72 pounds per square inch (101,325 newtons per sq. m). Commercial production of ammonia by the "Haber process" began in 1913, the year before World War I started. By the end of the war, in 1918, 200,000 tonnes of synthetic ammonia were being produced annually. Thanks to Haber, Germany had remained virtually self-sufficient in its supply of this vital chemical throughout the years of fighting.

WEAPONS OF WAR

In 1911 Haber became director of the Kaiser Wilhelm Institute for Physical Chemistry in Berlin. This was an important scientific center, and on the outbreak of World War I in August 1914 Haber put his laboratory at the disposal of the German government. He became a key figure in the development of poison gas as a weapon on the battlefield.

In 1918 Haber was awarded the Nobel Prize for Chemistry "for the synthesis of ammonia from its elements, nitrogen and hydrogen." The process that Haber had invented had produced the weapons responsible for the death and disablement of thousands of people in the war, so it seemed to many observers that this was an extremely odd award to make at this time. Many French and British scientists felt that the role Haber had personally played in

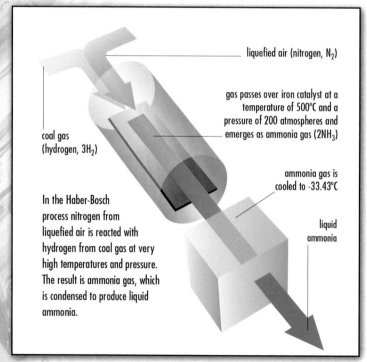

liquefied air (nitrogen, N_2)

gas passes over iron catalyst at a temperature of 500°C and a pressure of 200 atmospheres and emerges as ammonia gas ($2NH_3$)

coal gas (hydrogen, $3H_2$)

ammonia gas is cooled to -33.43°C

liquid ammonia

In the Haber-Bosch process nitrogen from liquefied air is reacted with hydrogen from coal gas at very high temperatures and pressure. The result is ammonia gas, which is condensed to produce liquid ammonia.

developing German chemical warfare should have made him unfit to receive such an award. The British scientific journal *Nature* commented that, in introducing chlorine gas, Haber had "initiated a mode of warfare which is to the everlasting discredit of Germany."

SEAWATER GOLD?

Ever the patriot, Haber sought another way to serve his country. At the end of World War I the Allies ordered Germany to pay financial reparations (a sort of fine) as a punishment. Some of this was to be paid in gold. The Allies hoped that the burden of such costs would halt any further military aggression by Germany.

Haber knew that the world's oceans contained billions of tons of gold, and he began to ponder how a chemist could recover this gold from seawater and so help free Germany from its huge debts. In 1923 he and his team set up their laboratory on the liner *Hansa* and began trying to extract gold from the Atlantic Ocean as they sailed to New York. However, on this and later voyages the amounts of gold they managed to recover were so tiny that Haber eventually abandoned the project in 1927.

THE FALL OF GERMAN SCIENCE

After the war, Haber's institute in Berlin continued to gain in importance, becoming the world's leading center of research in physical chemistry. Haber had close links with industry, and he also began building better relations with foreign scientists. In 1930 he founded the Japan Institute, which had bases in Berlin and Tokyo, to forge closer cultural and scientific links between the two countries.

In January 1933, Adolf Hitler (1889–1945) became chancellor of Germany. The Nazi Party, which he headed, blamed the Jews for the country's economic and social problems and passed a series of anti-Jewish laws. In April 1933, a new Civil Service Law came into force, preventing the employment of anyone who was not of pure German descent. All university teachers and professors in Germany were employed by the state, and very large numbers of them, particularly scientists and doctors, were Jewish. Hitler stated baldly that if the loss of Jewish scientists meant the "annihilation of German science, then we shall do without science for a few years."

For Jewish scientists, however distinguished or old, leaving the country was the only sensible course of action, even though the costs of doing so were high. It meant abandoning one's family home, all financial assets, and perhaps a laboratory or a professional position built up over a lifetime. Non-Jewish German scientists looked on hopelessly as they saw the might of German science dwindle away before their eyes. Nobel Prize winners, major mathematicians, physicists, and chemists were all forced to flee their own country.

Haber was Jewish. At first his leading position, famous patriotism, and

On the outbreak of World War I in 1914, Haber had put his laboratory at the disposal of the German government, but his past loyalty did not count for anything under Hitler's regime. In 1933, "the Jew Haber" resigned his position and left Germany for the last time.

previous service to the state gave him some protection. Those, like Haber, who had fought in the 1914–18 war were exempt from the Civil Service Law. Unofficially, Haber was told that—as long as he made no fuss—his position would be secure. Haber would have none of this, though. He resigned his post and told the ministry: "For more than 40 years I have selected my collaborators on the basis of their intelligence and their character and not on the basis of their grandmothers, and I am not willing for the rest of my life to change this method which I have found to work so well."

There was no turning back. Haber left Germany for the final time in the summer of 1933. He had already been invited to work in Cambridge, England, by Sir William Pope (1870–1939), the chemist who had been in charge of Britain's own chemical warfare program. But Haber was to work in Cambridge for only a brief period. Soon after his move to England, while on vacation in Switzerland on January 29, 1934, he died of a heart attack.

SCIENTIFIC BACKGROUND

POLITICAL AND CULTURAL BACKGROUND

Before 1885

French chemist Claude-Louis Berthollet (1748–1822) shows that ammonia is a compound of nitrogen and hydrogen

German chemist August Wilhelm von Hofmann (1818–1892) produces synthetic dyes known as magenta, or fuchsine, and "Hofmann's violets"

1888 Wilhelm II (1859–1941) is crowned third German emperor and ninth king of Prussia

1889 At the Hague conference, Germany and 23 other countries agree not to use chemical weapons: "projectiles whose sole object is the diffusion of asphyxiating gases"

1890

1893 Haber begins studies on decomposition of organic compounds at high temperatures

1895

1896 Haber publishes *Experimental Studies on Decomposition of Hydrocarbons*

1900

1900 The Republican president of the United States, William McKinley (1843–1901), wins re-election on the promise of "four more years of the full dinner pail"

1902 Scottish-born author Arthur Conan Doyle (1859–1930) publishes *The Hound of the Baskervilles*, featuring his popular fictional detective, Sherlock Holmes

1905 German chemist Johann von Baeyer (1835–1917) wins the Nobel Prize for chemistry for synthesizing indigo

1906 English chemist William Henry Perkin (1838–1907) is honored at a celebration dinner at the Royal Institution in London; it is the 50th anniversary of his discovery of the synthetic dye, mauve

1905 Russian troops in St. Petersburg open fire on workers demonstrating in front of the Winter Palace of Tsar Nicholas II (1868–1918); more than 200 people are killed

1905

1909 The first 3.5 ounces (100 g) of synthetic ammonia are produced by the Haber process

1910

1909 The National Negro Committee (later the National Association for the Advancement of Colored People [NAACP]) is founded in the United States

1912 German industrial chemist Friedrich Karl Rudolf Bergius (1884–1949) uses high pressure, high temperatures, and a catalyst to produce paraffins (alkanes) such as petrol and kerosene from coal dust

1913 The Haber process is developed for commercial production of nitrogen by German chemist Karl Bosch (1874–1940) at Badische Anilin- and Soda-Fabrik in Oppau, Germany

1915 In April at Ypres in Belgium, German soldiers attack French and Canadian troops using chlorine gas; more than 5,000 Allied troops die

1915

1916 In France, the Germans and French each lose 400,000 dead or wounded in the Battle of Verdun

1920

1922 The Union of Soviet Socialist Republics (USSR) is officially established

1924 American inventor Clarence Birdseye (1886–1956), who has seen how the people of Labrador keep food fresh by freezing it, sets up the General Seafood Company selling a range of frozen foods

c. 1925 Commercial production of paraffins from coal dust using Bergius's process begins; it will provide important alternative fuel for Germany during World War II (1939–45)

1925

1925 Bosch invents process for preparing hydrogen on a commercial scale

1928 In the Soviet Union Communist leader Josef Stalin (1879–1953) tries to force peasants to work state-owned farms, or "collectives," as part of his first Five-Year Plan; those who resist are killed or sent to labor camps

1930

After 1930

1939 Swiss chemist Paul Müller (1899–1965) synthesizes dichlorodiphenyl-trichloroethane (DDT) as an insecticide

1962 *Silent Spring*, by American naturalist Rachel Carson (1907–1964) highlights the devastating effect of modern synthetic chemical pesticides such as DDT on the food chain

PERIODIC TABLE OF ELEMENTS

The periodic table organizes all the chemical elements into a simple chart according to the physical and chemical properties of their atoms. The elements are arranged by atomic number from 1 to 118. The atomic number is based on the number of protons in the nucleus of the atom. The atomic mass is the combined mass of protons and neutrons in the nucleus. Each element has a chemical symbol that is an abbreviation of its name. In some cases, such as potassium,

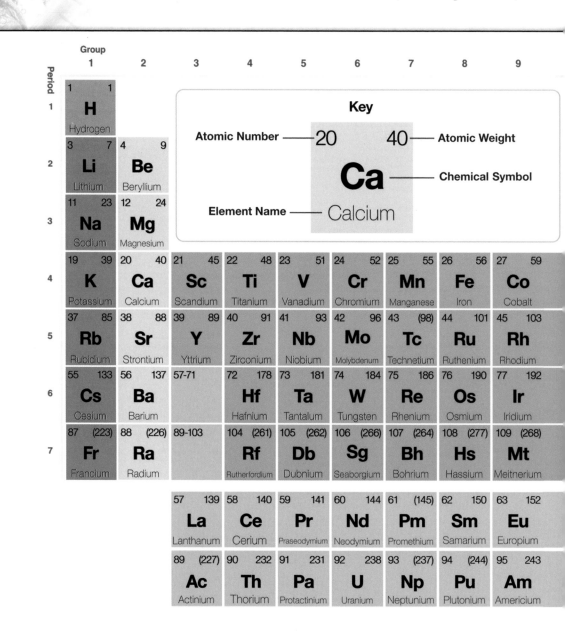

the symbol is an abbreviation of its Latin name ("K" stands for kalium). The name by which the element is commonly known is given in full underneath the symbol. The last item in the element box is the atomic mass. This is the average mass of an atom of the element.

Scientists have arranged the elements into vertical columns called groups and horizontal rows called periods. Elements in any one group all have the same number of electrons in their outer shell and have similar chemical properties. Periods represent the increasing number of electrons it takes to fill the inner and outer shells and become stable. When all the spaces have been filled (Group 18 atoms have all their shells filled) the next period begins.

10	11	12	13	14	15	16	17	18
								2 4 **He** Helium
			5 11 **B** Boron	6 12 **C** Carbon	7 14 **N** Nitrogen	8 16 **O** Oxygen	9 19 **F** Fluorine	10 20 **Ne** Neon
			13 27 **Al** Aluminum	14 28 **Si** Silicon	15 31 **P** Phosphorus	16 32 **S** Sulfur	17 35 **Cl** Chlorine	18 40 **Ar** Argon
28 59 **Ni** Nickel	29 64 **Cu** Copper	30 65 **Zn** Zinc	31 70 **Ga** Gallium	32 73 **Ge** Germanium	33 75 **As** Arsenic	34 79 **Se** Selenium	35 80 **Br** Bromine	36 84 **Kr** Krypton
46 106 **Pd** Palladium	47 108 **Ag** Silver	48 112 **Cd** Cadmium	49 115 **In** Indium	50 119 **Sn** Tin	51 122 **Sb** Antimony	52 128 **Te** Tellurium	53 127 **I** Iodine	54 131 **Xe** Xenon
78 195 **Pt** Platinum	79 197 **Au** Gold	80 201 **Hg** Mercury	81 204 **Tl** Thallium	82 207 **Pb** Lead	83 209 **Bi** Bismuth	84 (209) **Po** Polonium	85 (210) **At** Astatine	86 (222) **Rn** Radon
110 (281) **Ds** Darmstadtium	111 (280) **Rg** Roentgenium	112 (285) **Cn** Copernicium	113 (284) **Uut** Ununtrium	114 (289) **Fl** Flerovium	115 (288) **Uup** Ununpentium	116 (293) **Lv** Livermorium	117 (294) **Uus** Ununseptium	118 (294) **Uuo** Ununoctium

64 157 **Gd** Gadolinium	65 159 **Tb** Terbium	66 163 **Dy** Dysprosium	67 165 **Ho** Holmium	68 167 **Er** Erbium	69 169 **Tm** Thulium	70 173 **Yb** Ytterbium	71 175 **Lu** Lutetium
96 (247) **Cm** Curium	97 (247) **Bk** Berkelium	98 (251) **Cf** Californium	99 (252) **Es** Einsteinium	100 (257) **Fm** Fermium	101 (258) **Md** Mendelevium	102 (259) **No** Nobelium	103 (262) **Lr** Lawrencium

acid Substance that dissolves in water to form hydrogen ions (H+). Acids are neutralized by alkalis and have a pH below 7.

activation energy The energy needed for reactants to change into products in a chemical reaction.

alchemist Person who attempts to change one substance into another using a combination of primitive chemistry and magic.

alkali Substance that dissolves in water to form hydroxide ions (OH-). Alkalis have a pH greater than 7 and will react with acids to form salts.

allotrope A different form of an element in which the atoms are arranged in a different structure.

alloy A metallic substance that contains two or more metals. An alloy may also be made of a metal and a small amount of a nonmetal. Steel is an alloy of iron and carbon.

alpha particle The nucleus of a helium atom. This particle has two protons and two neutrons.

atom The smallest independent building block of matter. All substances are made of atoms.

atomic mass number The number of protons and neutrons in an atom's nucleus.

atomic number The number of protons in a nucleus.

boiling point The temperature at which a liquid turns into a gas.

bond The chemical connection between atoms.

by-product A substance that is produced when another material is made.

calorimeter Apparatus for accurately measuring heat given out or taken in.

catalyst Substance that speeds up a chemical reaction but is left unchanged at the end of the reaction.

chemical equation Symbols and numbers that show how reactants change into products during a chemical reaction.

chemical formula The letters and numbers that represent a chemical compound, such as "H_2O" for water.

chemical reaction The reaction of two or more chemicals (the reactants) to form new chemicals (the products).

chemical symbol The letters that represent a chemical, such as "Cl" for chlorine or "Na" for sodium.

combination reaction A reaction in which two or more reactants combine to form one product.

combustion The reaction that causes burning. Combustion is generally a reaction with oxygen in the air.

compound Substance made from more than one element and that has undergone a chemical reaction.

compress To reduce in size or volume by squeezing or exerting pressure.

condensation The change of state from a gas to a liquid.

condensation reaction A reaction between two molecules that produces water as a by-product.

conductor A substance that carries electricity and heat well.

corrosion The slow wearing away of metals or solids by chemical attack.

covalent bond Bond in which atoms share electrons.

crystal A solid made of regular repeating patterns of atoms.

crystal lattice The arrangement of atoms in a crystalline solid.

decay chain The sequence in which a radioactive element breaks down.

decomposition reaction A reaction in which a single compound breaks down into two or more substances.

density The mass of substance in a unit of volume.

dipole attraction The attractive force between the electrically charged ends of molecules.

displacement reaction A reaction that occurs when a more reactive atom replaces a less reactive atom in a compound.

dissolve To form a solution.

ductile Describes materials that can be stretched into a thin wire. Many metals are ductile.

elastic Describes a substance that returns to its original shape after being stretched.

electricity A stream of electrons or other charged particles moving through a substance.

electrolyte Liquid containing ions that carries a current between electrodes.

electron A tiny negatively charged particle that moves around the nucleus of an atom.

electronegativity The power of an atom to attract an electron. Nonmetals, which have only a few spaces in their outer shell, are the most electronegative. Metals, which have several empty spaces in their outer shell, tend to lose electrons in chemical reactions. Metals of this type are termed electropositive.

element A material that cannot be broken up into simpler ingredients. Elements contain only one type of atom.

endothermic reaction A reaction that absorbs heat energy.

energy The ability to do work.

energy level Electron shells represent different energy levels. Those closest to the nucleus have the lowest energy.

enthalpy The change in energy during a chemical reaction.

evaporation The change of state from a liquid to a gas when the liquid is at a temperature below its boiling point.

exothermic reaction A reaction that releases energy.

fission Process by which a large atom breaks up into two or more smaller fragments.

four elements The ancient theory that all matter consisted of only four elements (earth, air, fire, and water) and their combinations.

fusion When small atoms fuse to make a single larger atom.

gaseous state State in which particles are not joined and are free to move in any direction.

heat The transfer of energy between atoms. Adding heat makes atoms move more quickly.

hydrocarbon Chemical compounds that contain only hydrogen and carbon.

hydrogen bond A weak dipole attraction that always involves a hydrogen atom.

hydrolysis The process by which a molecule splits after reacting with a molecule of water.

hydroxyl A functional group (–OH) made up of an oxygen atom and a hydrogen atom.

inhibitor A substance that slows down a chemical reaction without being used up by it; also called a negative catalyst.

insoluble A substance that does not dissolve in a solvent.

insulator A substance that does not transfer an electric current or heat.

intermolecular bonds The bonds that hold molecules together. These bonds are weaker than those between atoms in a molecule.

intramolecular bond Strong bond between atoms in a molecule.

ion An atom that has lost or gained one or more electrons.

ionic bond Bond in which one atom gives one or more electrons to another atom.

ionization The formation of ions by adding or removing electrons from atoms.

isotope Atoms of a given element must have the same number of protons but can have different numbers of neutrons. These different versions of the same element are called isotopes.

liquid Substance in which particles are loosely bonded and are able to move freely around each other.

matter Anything that can be weighed.

melting point The temperature at which a solid changes into a liquid. When a liquid changes into a solid, this same temperature is also called the freezing point.

metallic bond Bond in which outer electrons are free to move in the spaces between the atoms.

mixture Matter made from different types of substances that are not physically or chemically bonded together.

mole The amount of any substance that contains the same number of atoms as in 12 grams of carbon-12 atoms. This number is 6.022×10^{23}.

molecule Two or more bonded atoms that form a substance with specific properties.

neutralization reaction A displacement reaction involving acids and bases in which one of the products is water and the other is a salt.

neutron One of the particles that make up the nucleus of an atom. Neutrons do not have any electric charge.

nucleus The central part of an atom. The nucleus contains protons and neutrons. The exception is hydrogen, in which the nucleus contains only one proton.

octet rule If an atom's outer electron shell contains exactly eight atoms (an octet), the atom will be stable and unreactive.

oxidation The addition of oxygen to a compound.

oxidation state A number used to describe how many electrons an atom can lose or gain.

oxidizer A substance that removes electrons from another substance to make its own outer shell stable.

pH A measure of the acidity or alkalinity of a substance.

photon A particle that carries a quantity of energy, such as in the form of light.

potential energy Energy that something has because of the way it is positioned or its parts are arranged.

product The new substance created by a chemical reaction.

proton A positively charged particle in an atom's nucleus.

radiation The products of radioactivity—alpha and beta particles and gamma rays.

radioactive decay The breakdown of an unstable nucleus through the loss of alpha and beta particles.

reactants The ingredients necessary for a chemical reaction.

redox reaction A reaction that occurs when electrons move from one of the reactants to the other.

relative atomic mass A measure of the mass of an atom compared with the mass of another atom. The values used are the same as those for atomic mass.

relative molecular mass The sum of all the atomic masses of the atoms in a molecule.

salt A compound made from positive and negative ions that forms when an alkali reacts with an acid.

shell The orbit of an electron. Each shell can contain a specific number of electrons and no more.

solid State of matter in which particles are held in a rigid arrangement.

solute A substance that dissolves in a solvent.

solution A mixture of two or more elements or compounds in a single state (solid, liquid, or gas).

solvent The liquid that dissolves a solute.

standard conditions Normal room temperature and pressure.

state The form that matter takes—either a solid, a liquid, or a gas.

subatomic particles Particles that are smaller than an atom.

temperature A measure of how fast molecules are moving.

valence electrons The electrons in the outer shell of an atom.

van der Waals forces Short-lived forces between atoms and molecules.

volatile Describes a liquid that evaporates easily.

voltage The force that pushes electrons through an electric circuit.

American Association for the
Advancement of Science
1200 New York Avenue NW
Washington, DC 20005
(202) 326-6400
Web site: http://www.aaas.org
An international non-profit organization
dedicated to advancing science
around the world by serving as an
educator, leader, spokesperson, and
professional association.

American Chemical Society
1155 Sixteenth Street NW
Washington, DC 20036
(800) 227-5558
Web site: http://www.acs.org
ACS is a congressionally chartered independent membership organization
that represents professionals at all
degree levels and in all fields of
chemistry and sciences that involve
chemistry.

American Institute of Chemical
Engineers
3 Park Avenue 19 Floor
New York, NY 10016-5991
(800) 242-4363
Web site: http://www.aiche.org
The world's leading organization for
chemical engineering professionals,
with more than 45,000 members
from over 90 countries.

Chemical Heritage Foundation
315 Chestnut Street
Philadelphia, PA 19106
(215) 925-2222

Web site: http://www.chemheritage.org
A collections-based nonprofit organization that preserves the history and
heritage of chemistry, chemical engineering, and related sciences and
technologies.

International Union of Pure and Applied
Chemistry
P.O. Box 13757
Research Triangle Park, NC 27709-3757
Web site: http://www.iupac.org
Serves to advance the worldwide aspects
of the chemical sciences and to con-
tribute to the application of
chemistry in the service of mankind.

Museum of Science and Industry
5700 S. Lake Shore Drive
Chicago, IL 60637
(773) 684-1414
Web site: http://www.msichicago.org
One of the largest science museums in
the world is home to more than
35,000 artifacts and nearly 14 acres
of hands-on exhibits designed to
spark scientific inquiry and
creativity.

WEB SITES

Due to the changing nature of Internet
links, Rosen Publishing has developed
an online list of Web sites related to the
subject of this book. This site is updated
regularly. Please use this link to access
the list:

http://www.rosenlinks.com/CORE/React

Ardley, Neil. *101 Great Science Experiments*. New York, NY: DK, 2006.

Basher, Simon. *Basher Science: Chemistry: Getting a Big Reaction*. London, England: Kingfisher, 2010.

Boothroyd, Jennifer. *Many Kinds of Matter: A Look at Solids, Liquids, and Gases*. Minneapolis, MN: Lerner, 2011.

Brent, Lynnette. *Acids and Bases*. New York, NY: Crabtree Publishing, 2008.

Brent, Lynnette. *Chemical Changes*. New York, NY: Crabtree Publishing, 2008.

Brown, Cynthia Light. *Amazing Kitchen Chemistry Projects You Can Build Yourself*. White River Junction, VT: Nomad Press, 2008.

Coelho, Alexa, and Simon Quellan Field. *Why Is Milk White? & 200 Other Curious Chemistry Questions*. Chicago, IL: Chicago Review Press, 2013.

Connolly, Sean. *The Book of Totally Irresponsible Science: 64 Daring Experiments for Young Scientists*. New York, NY: Workman, 2008.

Furgang, Adam. *The Noble Gases: Helium, Neon, Argon, Krypton, Xenon, Radon*. New York, NY: Rosen Publishing, 2010.

Hasan, Heather. *The Boron Elements: Boron, Aluminum, Gallium, Indium, Thallium*. New York, NY: Rosen Publishing, 2009.

Heos, Bridget. *The Alkaline Earth Metals: Beryllium, Magnesium, Calcium, Strontium, Barium, Radium*. New York, NY: Rosen Publishing, 2009.

La Bella, Laura. *The Oxygen Elements: Oxygen, Sulfur, Selenium, Tellurium, Polonium*. New York, NY: Rosen Publishing, 2010.

Lew, Kristi. *The 15 Lanthanides and the 15 Actinides*. New York, NY: Rosen Publishing, 2010.

Lew, Kristi. *The Alkali Metals: Lithium, Sodium, Potassium, Rubidium, Cesium, Francium*. New York, NY: Rosen Publishing, 2009.

Newmark, Ann. *DK Eyewitness Books: Chemistry*. New York, NY: DK, 2005.

Roza, Greg. *The Halogen Elements: Fluorine, Chlorine, Bromine, Iodine, Astantine*. New York, NY: Rosen Publishing, 2010.

Silverstein, Alvin, Virginia B. Silverstein, and Laura Silverstein Nunn. *Matter*. Minneapolis, MN: Twenty-First Century Books, 2008.

Spangler, Steve. *Naked Eggs and Flying Potatoes: Unforgettable Experiments That Make Science Fun*. Austin, TX: Greenleaf Books, 2010.

TIME for Kids Big Book of Science Experiments: A Step-by-Step Guide. New York, NY: TIME for Kids, 2011.

VanCleave, Janice. *Step-by-Step Science Experiments in Chemistry*. New York, NY: Rosen, 2012.

PHOTO CREDITS